2007 - 2018 © Frank Ho, Amanda Ho, All rights reserved.   www.homathchess.com

Mom! I Learn Multiplication Using Math-Chess-Puzzles Connection

棋谜式乘法

Frank Ho   Amanda Ho

何数棋谜 培训

Ho Math Chess Learning Centre

2007 - 2018 © Frank Ho, Amanda Ho, All rights reserved.　　www.homathchess.com

Table of Contents

Ho Math Chess　何数棋谜　妈！我会棋谜式乘法啦！
Mom! I Learn Multiplication Using Math-Chess-Puzzles Connection!

Student's Name _____ Date _____

2007 - 2018 © Frank Ho, Amanda Ho, All rights reserved.　　www.homathchess.com

2007 - 2018 © Frank Ho, Amanda Ho, All rights reserved. www.homathchess.com

2007 - 2018 © Frank Ho, Amanda Ho, All rights reserved.　　www.homathchess.com

Preface 2017

I have added one digit divided by one digit division in this new updated multiplication workbook. The reason is that it does not matter how we train math teachers to ask students every time before the class starts on what they are learning at day school math class, sometimes the teacher forgot or the students were not able to communicate well, so one thing happened was the students already were doing 3-digit times 2-digit by using this workbook, but they have never done any divisions while their day schools start to do one-digit divided by one-digit. We do not want the students to carry too many workbooks, so this division addition serves as a transition before purchasing a division workbook.

2007 - 2018 © Frank Ho, Amanda Ho, All rights reserved.　　www.homathchess.com

About Ho Math Chess™ Math Workbooks

I have taught students from grade 1 to grade 12 since I opened the Vancouver **Ho Math Chess** Learning Centre in 1995. I have personally witnessed on how some students suffered because they could not master some very basic computational skills. I do not want to create a workbook, which is about practice, practice, and more practice of computational skills. This has motivated me to create a workbook that would be very different from the conventional ones in terms of the way the questions are presented to the students. I wanted students to learn basic computation skills by using the carefully designed worksheets so that students can master basic computation skills in an intuitive way. These worksheets were being designed while I actually watched student's work and modified accordingly to their responses.

I had an idea to create a computational workbook, which integrates chess knowledge, puzzles, and math in such a way that students could learn how to transfer abstract symbols into numerical values and then calculating the results by using puzzles-like problems. This idea may sound very simple but the result is much more profound – not only students learn to do math in multi-step, they also learn how to process information by converting abstract symbols into numerical values, which is important in learning critical thinking skill.

One very noticeable computation format in **Ho Math Chess** math workbooks is the way computation directions are presented - it is no longer just a linear fashion; instead, students work on computations in all kinds of directions: top down, bottom up, left to eight, right to left , diagonally, and even circular motion. For example, the multiplication workbook computation format is designed in such a way that it takes the boredom out by using the format of multi-direction computation and multi-concept learning. Students could be introduced division computation procedure while working on multiplication and even equivalent fractions but without realizing that they are actually working on advanced math concepts and mechanic computation procedure beyond their grade level. One other example is that the factoring procedure is introduced while students are working on multiplication. These many embedded computational procedures included in the elementary level of math workbook will benefit students when they go to higher grades.

Ho Math Chess 何数棋谜 妈！我会棋谜式乘法啦！
Mom! I Learn Multiplication Using Math-Chess-Puzzles Connection!

Student's Name _____ Date _____

2007 - 2018 © Frank Ho, Amanda Ho, All rights reserved.   www.homathchess.com

My idea of using multi-direction, multi-operation, multi-procedure, multi-concept learning style is the very distinct and innovative way of creating these workbooks. Students found them less boring and even willing to do the same worksheets the second time if they did not master the first time.

I am hoping by working through these addition, subtraction, multiplication workbooks, the division would be just a matter of fine-tuning its computation procedure.

In 2014, all computation workbooks have taken major upgrades to include truly math and chess integrated material, this idea is world first and these worksheets formats are also world first. With these releases of many new and innovative workbooks, the math teaching and tutoring has taken the entire math tutoring to a revolutionary stage. Because of the creation of integrated math, chess, and puzzles integrated workbooks, Ho Math Chess has made the dream of fun math teaching becomes true.

Students at Ho Math Chess have enjoyed math even more than the previous workbooks and we see dramatic changes in student's attitude, they are happier and more willing to work on math.

Frank Ho
Amanda Ho

July, 2014

In 2015, we added a new part called intelligent worksheets which allow students to figure out the operator and this is an innovative idea in computation format because all other math workbooks on the market all pre-define their operators for students and students just calculate the results.

November, 2015
Frank Ho
Amanda Ho

Chess pieces and their mathematical values

Symbols of chess pieces	Names of chess pieces	Mathematical values
	Queen	9
	Rook	5
	Bishop	3
	Knight	3
	Pawn	1
	King	0

Ho Math Chess 何数棋谜 妈!我会棋谜式乘法啦!

Mom! I Learn Multiplication Using Math-Chess-Puzzles Connection!

Student's Name _____ Date _____

2007 - 2018 © Frank Ho, Amanda Ho, All rights reserved. www.homathchess.com

From addition to multiplication

Fill in each ____ with a mathematical expression.

Diagram	Explanation	Addition	Multiplication	Comments
♟♟ ♟♟ ♟♟	3 groups of ♟♟	2 + 2 + 2 = 6	3 × 2 = 6	**3 times 2 is** (equals) **6.** 3 multiplied by 2 is (equals) 6. The product of the factors 3 and 2 is (equals) 6. Three 2s equal 6. 3 sets of 2 horses are 6 horses. Factor 3 times factor 2 is product 6. Multiplicand 3 times multiplier 2 is product 6.
♟♟ ♟♟ ♟♟ ♟♟	4 groups of ♟♟.	2 + 2 + 2 + 2 = 8	_____	
♟♟♟♟ ♟♟♟♟	2 groups of ♟♟♟♟.	4 + 4 = ___	_____	
♟♟♟♟ ♟♟♟♟ ♟♟♟♟ ♟♟♟♟	4 groups of ♟♟♟♟	_____	_____	
♟♟♟ ♟♟♟ ♟♟♟ ♟♟♟	4 groups of 3 ♟♟♟	_____	_____	

2007 - 2018 © Frank Ho, Amanda Ho, All rights reserved.   www.homathchess.com

Multiplication table

×	0	1	2	3	4	5	6	7	8	9
1	0	1	2	3	4	5	6	7	8	9
2	0	2	4	6	8	10	12	14	16	18
3	0	3	6	9	12	15	18	21	24	27
4	0	4	8	12	16	20	24	28	32	36
5	0	5	10	15	20	25	30	35	40	45
6	0	6	12	18	24	30	36	42	48	54
7	0	7	14	21	28	35	42	49	56	63
8	0	8	16	24	32	40	48	56	64	72
9	0	9	18	27	36	45	54	63	72	81

Student's Name _____ Date _____

2007 - 2018 © Frank Ho, Amanda Ho, All rights reserved.   www.homathchess.com

Multiplying with 0 and 1

2	♙	1	0	2
X 1	X 3	X 4	X ♗	X ♛
□	□	□	□	□

5	1	7	9	8
X ♙	X 5	X ♙	X 1	X 1
□	□	□	□	□

2	9	♚	1	2
X ♚	X 0	X 2	X 3	X 0
□	□	□	□	□

♙	8	4	1	♚
X 7	X 1	X 0	X 8	X 5
□	□	□	□	□

Ho Math Chess	何数棋谜　妈！我会棋谜式乘法啦！

Mom! I Learn Multiplication Using Math-Chess-Puzzles Connection!

Student's Name _____ Date _____

2007 - 2018 © Frank Ho, Amanda Ho, All rights reserved.　　www.homathchess.com

Counting 2's multiples (Doubling)

Place an ○ over every multiple of 2.

1	2	3	4	5	6	7	8	9	10
11	12	13	14	15	16	17	18	19	20
21	22	23	24	25	26	27	28	29	30
31	32	33	34	35	36	37	38	39	40
41	42	43	44	45	46	47	48	49	50
51	52	53	54	55	56	57	58	59	60
61	62	63	64	65	66	67	68	69	70
71	72	73	74	75	76	77	78	79	80
81	82	83	84	85	86	87	88	89	90
91	92	93	94	95	96	97	98	99	100

What are the unit digits of 2's multiples? _____.

Mom! I Learn Multiplication Using Math-Chess-Puzzles Connection!

Student's Name _____ Date _____

2007 - 2018 © Frank Ho, Amanda Ho, All rights reserved.   www.homathchess.com

Counting 2's multiples

Circle the following 2's multiples.

1 2 3 4 5 6 7 8 9 10 11 12 13 14 15 16 17 18

Fill in the following ☐ with a number.

Sequence	1	2	3	4	5	6	7	8	9
Add 2	☐	4	6	☐	10	12	☐	☐	18

Sequence	♙	2	♝	4	♜	6	7	8	♛
Add 2	2	☐	6	☐	10	☐	☐	☐	18

Sequence	1	2	3	4	5	6	7	8	9
Add 2	2	☐	6	☐	10	☐	14	☐	18

Sequence	♙	2	♝	4	♜	6	7	8	♛
Add 2	☐	4	☐	8	☐	12	☐	16	☐

Sequence	1	2	3	4	5	6	7	8	9
Add 2	☐	☐	☐	8	☐	12	☐	16	☐

Sequence	♙	2	♝	4	♜	6	7	8	♛
Add 2	☐	4	☐	☐	☐	12	☐	16	☐

2007 - 2018 © Frank Ho, Amanda Ho, All rights reserved.   www.homathchess.com

2 times

Fill in each _____ with multiplication expression.

$2 \times 1 =$	♟ ♟ =	$= 1 \times 2 =$	♟ ♟
$2 \times 2 =$	♟ ♟ ♟ ♟ =		♟ ♟ ♟ ♟
$2 \times 3 =$	♟ ♟ ♟ = ♟ ♟ ♟	$=$ _____ $=$	♟ ♟ ♟ ♟ ♟ ♟
$2 \times 4 =$	♟ ♟ ♟ ♟ = ♟ ♟ ♟ ♟	$= 4 \times 2 =$	♟ ♟ ♟ ♟ ♟ ♟ ♟ ♟
$2 \times 5 =$	♟ ♟ ♟ ♟ ♟ = ♟ ♟ ♟ ♟ ♟	$=$ _____ $=$	♟ ♟ ♟ ♟ ♟ ♟ ♟ ♟ ♟ ♟
$2 \times 6 =$	♟ ♟ ♟ ♟ ♟ ♟ = ♟ ♟ ♟ ♟ ♟ ♟	$= 6 \times 2 =$	♟ ♟ ♟ ♟ ♟ ♟ ♟ ♟ ♟ ♟ ♟ ♟
$2 \times 7 =$	♟ ♟ ♟ ♟ ♟ ♟ ♟ = ♟ ♟ ♟ ♟ ♟ ♟ ♟	$=$ _____ $=$	♟ ♟ ♟ ♟ ♟ ♟ ♟ ♟ ♟ ♟ ♟ ♟ ♟ ♟
$2 \times 8 =$	♟ ♟ ♟ ♟ ♟ ♟ ♟ ♟ = ♟ ♟ ♟ ♟ ♟ ♟ ♟ ♟	$= 8 \times 2 =$	♟ ♟ ♟ ♟ ♟ ♟ ♟ ♟ ♟ ♟ ♟ ♟ ♟ ♟ ♟ ♟

2 times

$1 + 1 = \square$	$= 2 \times 1 =$	$1 \times 2 = \square$	$\begin{array}{r} ♙ \\ + ♙ \\ \hline \end{array}$ $\square = 2 \times 1 = \square$
$2 + 2 = \square$	$= 2 \times 2 =$	$2 \times 2 = \square$	$\begin{array}{r} 2 \\ + 2 \\ \hline \end{array}$ $\square = 2 \times 2 = \square$
$3 + 3 = \square$	$= 2 \times 3 =$	$3 \times 2 = \square$	$\begin{array}{r} ♞ \\ + ♞ \\ \hline \end{array}$ $\square = 2 \times 3 = \square$
$4 + 4 = \square$	$= 2 \times 4 =$	$4 \times 2 =$ $\square$	$\begin{array}{r} 4 \\ + 4 \\ \hline \end{array}$ $\square = 2 \times 4 = \square$

Mom! I Learn Multiplication Using Math-Chess-Puzzles Connection!

Student's Name _____ Date _____

2007 - 2018 © Frank Ho, Amanda Ho, All rights reserved.   www.homathchess.com

$5 + 5 =$	☐	$= 2 \times$ ♖ $=$	$5 \times 2 =$	☐	$\begin{array}{r}\text{⬚}\\ +\,\text{⬚}\\\hline\end{array}$ ☐ $= 2 \times 5 =$ ☐
$6 + 6 =$	☐	$= 2 \times 6 =$	$6 \times 2 =$	☐	$\begin{array}{r}6\\ +\,6\\\hline\end{array}$ ☐ $= 2 \times 6 =$ ☐
$7 + 7 =$	☐	$= 2 \times 7 =$	$7 \times 2 =$	☐	$\begin{array}{r}7\\ +\,7\\\hline\end{array}$ ☐ $= 2 \times 7 =$ ☐
$8 + 8 =$	☐	$= 2 \times 8 =$	$8 \times 2 =$	☐	$\begin{array}{r}8\\ +\,8\\\hline\end{array}$ ☐ $= 2 \times 8 =$ ☐
$9 + 9 =$	☐	$= 2 \times 9 =$	$9 \times 2 =$	☐	$\begin{array}{r}♕\\ +\,♕\\\hline\end{array}$ ☐ $= 2 \times 9 =$ ☐

Student's Name _____ Date _____

2007 - 2018 © Frank Ho, Amanda Ho, All rights reserved.   www.homathchess.com

2 times

$2 \times 1 = \square$	Two times one is $\square$	$1 \times 2 = \square$	One times two is $\square$
$2 \times 2 = \square$	Two times two is $\square$	$2 \times 2 = \square$	Two times two is $\square$
$2 \times 3 = \square$	Two times three is $\square$	$3 \times 2 = \square$	Three times two is $\square$
$2 \times 4 = \square$	Two times four is $\square$	$4 \times 2 = \square$	Four times two is $\square$
$2 \times 5 = \square$	Two times five is $\square$	$5 \times 2 = \square$	Five times two is $\square$
$2 \times 6 = \square$	Two times six is $\square$	$6 \times 2 = \square$	Six times two is $\square$
$2 \times 7 = \square$	Two times seven is $\square$	$7 \times 2 = \square$	Seven times two is $\square$
$2 \times 8 = \square$	Two times eight is $\square$	$8 \times 2 = \square$	Eight times two is $\square$
$2 \times 9 = \square$	Two times nine is $\square$	$9 \times 2 = \square$	Nine times two is $\square$

$$2 \qquad ♙ \qquad 2 \qquad 2 \qquad 2$$
$$\times 1 \qquad \times 2 \qquad \times 2 \qquad \times 3 \qquad \times ♝$$

$$5 \qquad 2 \qquad 7 \qquad 2 \qquad 9$$
$$\times 2 \qquad \times 6 \qquad \times 2 \qquad \times 8 \qquad \times 2$$

$$7 \qquad 2 \qquad ♜ \qquad 2 \qquad 2$$
$$\times 2 \qquad \times 8 \qquad \times 2 \qquad \times 6 \qquad \times 9$$

Ho Math Chess 何数棋谜 妈！我会棋谜式乘法啦！
Mom! I Learn Multiplication Using Math-Chess-Puzzles Connection!

Student's Name _____ Date _____

2007 - 2018 © Frank Ho, Amanda Ho, All rights reserved.   www.homathchess.com

2	♙	4	2	2
X 1	X 2	X 2	X 4	X 3

2	♘	2	8	2
X 3	X 2	X 7	X 2	X 9

4	2	5	2	7
X 2	X 4	X 2	X ♗	X 2

2	9	3	3	♗
X ♘	X 2	X 2	X 2	X 2

2	2	2	2	2
X 3	X 6	X 7	X 3	X 9

2007 - 2018 © Frank Ho, Amanda Ho, All rights reserved.   www.homathchess.com

Oral practice

two one two	2 ♟ ☐	$\begin{array}{r} 1\,1 \\ \times\ \ 2 \\ \hline \square\square \end{array}$
two two four	2 2 ☐	$\begin{array}{r} 2\,2 \\ \times\ \ 2 \\ \hline \square\square \end{array}$
two three six	2 ♗ ☐	$\begin{array}{r} 3\,3 \\ \times\ \ 2 \\ \hline \square\square \end{array}$
two four eight	2 4 ☐	$\begin{array}{r} 4\,4 \\ \times\ \ 2 \\ \hline \square\square \end{array}$
two five ten	2 5 ☐	$\begin{array}{r} {}^{1}\ \ \ \\ 5\,5 \\ \times\ \ 2 \\ \hline \square\square\square \end{array}$

2007 - 2018 © Frank Ho, Amanda Ho, All rights reserved.　　www.homathchess.com

Oral practice

two six twelve	2 6 ☐	$\begin{array}{r}1\\ 6\,6\\ \times\quad 2\\ \hline \square\square\square\end{array}$
two seven fourteen	2 7 ☐	$\begin{array}{r}7\,7\\ \times\quad 2\\ \hline \square\square\square\end{array}$
two eight sixteen	2 8 ☐	$\begin{array}{r}8\,8\\ \times\quad 2\\ \hline \square\square\square\end{array}$
two nine eighteen	2 9 ☐	$\begin{array}{r}9\,9\\ \times\quad 2\\ \hline \square\square\square\end{array}$
two five ten	2 ♖ ☐	$\begin{array}{r}5\,5\\ \times\quad 2\\ \hline \square\square\square\end{array}$
two six twelve	2 6 ☐	$\begin{array}{r}6\,6\\ \times\quad 2\\ \hline \square\square\square\end{array}$

Ho Math Chess 何数棋谜 妈!我会棋谜式乘法啦!
Mom! I Learn Multiplication Using Math-Chess-Puzzles Connection!

Student's Name _____ Date _____

2007 - 2018 © Frank Ho, Amanda Ho, All rights reserved. www.homathchess.com

Fill in _____ and ☐ with answers.

Times	Grouping	Addition
2 × \$1 = ☐	2 of \$1 = \$2	\$1 + \$1= ☐
♟ × \$2 = ☐	1 of \$2 = \$2	2 of \$1= ☐
2 × \$2 = ☐	2 of ☐ = \$4	\$2 + \$2 = ☐

Fill in _____ and ☐ with answers.

Expression	Grouping	Addition
2 × \$1 =	2 of ☐ = \$2	\$♟ + \$♟ = ☐
1 × \$2 =	♟ of ☐ = \$2	\$2 + \$2 = ☐

2 × 1 = ☐ = ♟ × 2 = ☐	2 × 2 = ☐ = 2 × 2 = ☐
2 × ☐ = 2 = 1 × ☐ = ☐	2 × ☐ = 4 = 2 × ☐ = ☐

2 ♟ ☐	2 5 ☐	2 ♕ ☐	2 4 ☐	2 8 ☐
2 2 ☐	2 6 ☐	2 1 ☐	2 5 ☐	2 9 ☐
2 3 ☐	2 7 ☐	2 2 ☐	2 6 ☐	2 ♟ ☐
2 4 ☐	2 8 ☐	2 3 ☐	2 7 ☐	2 2 ☐

Student's Name _____ Date _____

2007 - 2018 © Frank Ho, Amanda Ho, All rights reserved.   www.homathchess.com

Fill in _____ and ☐ with answers.

Times	Grouping	Addition
$2 \times \$3 = \square$	2 of $\square = 6$	$\$3 + \$3 = \square$
$3 \times \$2 = \square$	3 of $\square = 6$	$\$2 + \$2 + \$2 = \square$

Fill in _____ and ☐ with answers.

Expression	Grouping	Addition
$2 \times \$4$	2 of $\square = \$8$	$\$4 + \$4 = \square$
$4 \times \$2$	4 of $\square = \$8$	$\$2 + \$2 + \$2 + \$2 = \square$

$2 \times 3 = \square = 3 \times 2 = \square$	$2 \times 4 = \square = 4 \times 2 = \square$
$2 \times \square = 6 = 3 \times \square = \square$	$2 \times \square = 8 = 4 \times \square = \square$

2 ♙ ☐	2 5 ☐	2 9 ☐	2 4 ☐	2 8 ☐
2 2 ☐	2 6 ☐	2 1 ☐	2 ♖ ☐	2 ♕ ☐
2 3 ☐	2 7 ☐	2 2 ☐	2 6 ☐	2 1 ☐
2 4 ☐	2 8 ☐	2 3 ☐	2 7 ☐	2 2 ☐

Mom! I Learn Multiplication Using Math-Chess-Puzzles Connection!

Student's Name _____ Date _____

2007 - 2018 © Frank Ho, Amanda Ho, All rights reserved.　　www.homathchess.com

Fill in _____ and ☐ with answers.

Times	Grouping	Addition
$2 \times \$5 = \square$	2 of $\square$ = 10	$\$5 + \$$ ♜ $= \square$
$5 \times \$2 = \square$	5 of $\square$ = 10	$\$2 + \$2 + \$2 + \$2 + \$2 = \square$

Times	Grouping	Addition
$2 \times \$6 = \square$	2 of $\square$ = \$12	$\$6 + \$6 = \square$
$6 \times \$2 = \square$	6 of $\square$ = \$12	$\$2 + \$2 + \$2 + \$2 + \$2 + \$2 = \square$

$2 \times$ ♜ $= \square = 5 \times 2 = \square$	$2 \times 6 = \square = 6 \times 2 = \square$
$2 \times \square = 10 = 5 \times \square = \square$	$2 \times \square = 12 = 6 \times \square = \square$

2 ♙ $\square$	2 5 $\square$	2 ♛ $\square$	2 4 $\square$	2 8 $\square$
2 2 $\square$	2 6 $\square$	2 1 $\square$	2 ♜ $\square$	2 9 $\square$
2 3 $\square$	2 7 $\square$	2 2 $\square$	2 6 $\square$	2 ♙ $\square$
2 4 $\square$	2 8 $\square$	2 3 $\square$	2 7 $\square$	2 2 $\square$

Ho Math Chess 何数棋谜 妈！我会棋谜式乘法啦！

Mom! I Learn Multiplication Using Math-Chess-Puzzles Connection!

Student's Name _____ Date _____

2007 - 2018 © Frank Ho, Amanda Ho, All rights reserved.   www.homathchess.com

Fill in _____ and ☐ with answers.

Times	Grouping	Addition
2 × $7 = ☐	2 of ☐ = 14	$7 + $ 7 = ☐
7 × $2 = ☐	7 of ☐ = 14	$2 + $2 + $2 + $2 + $2 + $2 + $2 = ☐

Times	Grouping	Addition
2 × $8 = ☐	2 of ☐ = $16	$8 + $8 = ☐
8 × $2 = ☐	8 of ☐ = $16	$2 + $2 + $2 + $2 + $2 + $2 + $2 + $2 = ☐

2 × 7 = ☐ = 7 × 2 = ☐	2 × 8 = ☐ = 8 × 2 = ☐
2 × ☐ = 14 = 7 × ☐ = ☐	2 × ☐ = 16 = 8 × ☐ = ☐

2 1 ☐	2 5 ☐	2 ♛ ☐	2 4 ☐	2 8 ☐
2 2 ☐	2 6 ☐	2 1 ☐	2 5 ☐	2 9 ☐
2 ♗ ☐	2 7 ☐	2 2 ☐	2 6 ☐	2 1 ☐
2 4 ☐	2 8 ☐	2 3 ☐	2 7 ☐	2 2 ☐

Student's Name _____ Date _____

2007 - 2018 © Frank Ho, Amanda Ho, All rights reserved.   www.homathchess.com

Fill in _____ and ☐ with answers.

Times	Grouping	Addition
$2 \times \$9 = $ ☐	2 of ☐ = 18	$\$♛ + \$♛ = $ ☐
$7 \times \$2 = $ ☐	7 of ☐ = 14	$\$2 + \$2 + \$2 + \$2 + \$2 + \$2 + \$2 = $ ☐

Times	Grouping	Addition
$2 \times \$8 = $ ☐	2 of ☐ = \$16	$\$8 + \$8 = $ ☐
$8 \times \$2 = $ ☐	8 of ☐ = \$16	$\$2 + \$2 + \$2 + \$2 + \$2 + \$2 + \$2 + \$2 = $ ☐

$2 \times 9 = $ ☐ $= ♛ \times 2 = $ ☐	$2 \times 8 = $ ☐ $= 8 \times 2 = $ ☐
$2 \times $ ☐ $= 18 = 9 \times $ ☐ $= $ ☐	$2 \times $ ☐ $= 16 = 8 \times $ ☐ $= $ ☐

2 1 ☐	2 5 ☐	2 9 ☐	2 4 ☐	2 8 ☐
2 2 ☐	2 6 ☐	2 ♙ ☐	2 ♖ ☐	2 ♛ ☐
2 ♞ ☐	2 7 ☐	2 2 ☐	2 6 ☐	2 1 ☐
2 4 ☐	2 8 ☐	2 ♞ ☐	2 7 ☐	2 2 ☐

Ho Math Chess 何数棋谜 妈！我会棋谜式乘法啦！

Mom! I Learn Multiplication Using Math-Chess-Puzzles Connection!

Student's Name _____ Date _____

2007 - 2018 © Frank Ho, Amanda Ho, All rights reserved.   www.homathchess.com

Preparing for division

☐	☐	☐	☐	☐
X 2	X 2	X 4	X 2	X 3
4	6	8	10	12

☐	☐	☐	☐	☐
X 2	X 6	X 2	X 3	X 2
6	12	14	6	18

☐	☐	☐	☐	☐
X 4	X 2	X ♖	X 2	X 7
8	16	10	8	14

☐	☐	☐	☐	☐
X 2	X 2	X 2	X ♗	X 2
6	18	8	6	16

☐	☐	☐	☐	☐
X 4	X 5	X 2	X 2	X 7
8	10	16	8	14

2007 - 2018 © Frank Ho, Amanda Ho, All rights reserved.　www.homathchess.com

Preparing for division

☐ X 2 = 2	X ☐ 2)‾2‾	☐)2 X 2
☐ X 2 = 4	X ☐ 2)‾4‾	☐)4 X 2
☐ X 2 = 6	X ☐ 2)‾6‾	☐)6 X 2
☐ X 2 = 8	X ☐ 2)‾8‾	☐)8 X 2
☐ X 2 = 10	X ☐ 2)‾10‾	☐)10 X 2
☐ X 2 = 12	X ☐ 2)‾12‾	☐)12 X 2
☐ X 2 = 14	X ☐ 2)‾14‾	☐)14 X 2

2007 - 2018 © Frank Ho, Amanda Ho, All rights reserved.   www.homathchess.com

Preparing for division

☐ X 2 = 16	X ☐ 2)‾16‾	☐) 16 X 2
☐ X 2 = 18	X ☐ 2)‾18‾	☐) 18 X 2
☐ X 2 = 2	X ☐ 2)‾2‾	☐) 2 X 2
☐ X 2 = 4	X ☐ 2)‾4‾	☐) 4 X 2
☐ X 2 = 6	X ☐ 2)‾6‾	☐) 6 X 2
☐ X 2 = 8	X ☐ 2)‾8‾	☐) 8 X 2
☐ X 2 = 10	X ☐ 2)‾10‾	☐) 10 X 2

Ho Math Chess   何数棋谜   妈！我会棋谜式乘法啦！
Mom! I Learn Multiplication Using Math-Chess-Puzzles Connection!
Student's Name _____   Date _____

2007 - 2018 © Frank Ho, Amanda Ho, All rights reserved.   www.homathchess.com

Cross multiplication

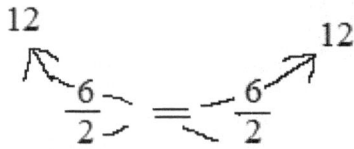

12   12

$$\frac{6}{2} = \frac{6}{2}$$

The left bottom 2 times the top right 6 is 12 by using the forward diagonal direction.

The right bottom 2 times the top left 6 is also 12 by using the backward diagonal direction.

12 12 $$\frac{6}{2} = \frac{6}{2}$$	☐ ☐ $$\frac{2}{2} = \frac{2}{2}$$	☐ ☐ $$\frac{2}{2} = \frac{3}{3}$$	☐ ☐ $$\frac{2}{2} = \frac{4}{4}$$
☐ ☐ $$\frac{2}{2} = \frac{5}{5}$$	☐ ☐ $$\frac{2}{2} = \frac{6}{6}$$	☐ ☐ $$\frac{2}{2} = \frac{7}{7}$$	☐ ☐ $$\frac{2}{2} = \frac{9}{9}$$
☐ ☐ $$\frac{2}{2} = \frac{5}{5}$$	☐ ☐ $$\frac{2}{2} = \frac{4}{4}$$	☐ ☐ $$\frac{2}{2} = \frac{7}{7}$$	☐ ☐ $$\frac{2}{2} = \frac{8}{8}$$

Ho Math Chess 何数棋谜 妈!我会棋谜式乘法啦!
Mom! I Learn Multiplication Using Math-Chess-Puzzles Connection!

Student's Name _____ Date _____

2007 - 2018 © Frank Ho, Amanda Ho, All rights reserved. www.homathchess.com

Different ways of writing multiplication (Learning division while doing multiplications)

$$2$$
$$\times$$
$$2$$

$$\frac{4}{\boxed{}\times} = 2 \qquad\qquad \frac{4}{\boxed{}\times} = 2$$

$$\nwarrow \; \| \; \nearrow$$

$$2 \times \boxed{} \; = \; \boxed{} \; = \; 2 \times \boxed{}$$

$$\swarrow \; \| \; \searrow$$

$$\times \boxed{} \qquad\qquad 2 \qquad\qquad \times \boxed{}$$

$$2\overline{)4} \qquad\qquad \times \qquad\qquad 2\overline{)4}$$
$$\boxed{}$$

$$\| \qquad\qquad\qquad\qquad \|$$

$$2\overline{)4} \qquad\qquad\qquad\qquad 2\overline{)4}$$
$$\times \boxed{} \qquad\qquad\qquad \times \boxed{}$$

30

2007 - 2018 © Frank Ho, Amanda Ho, All rights reserved.   www.homathchess.com

Different ways of writing multiplication (Learning division while doing multiplications)

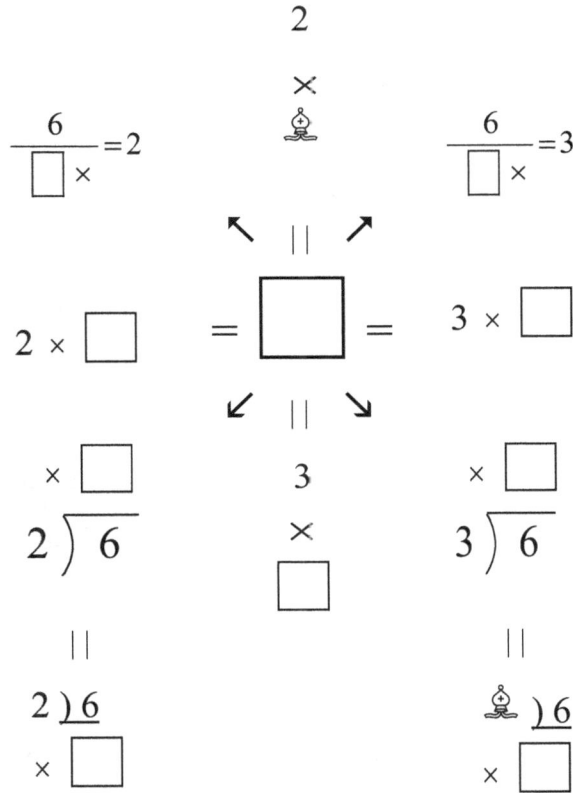

2007 - 2018 © Frank Ho, Amanda Ho, All rights reserved.   www.homathchess.com

Different ways of writing multiplication (Learning division while doing multiplications)

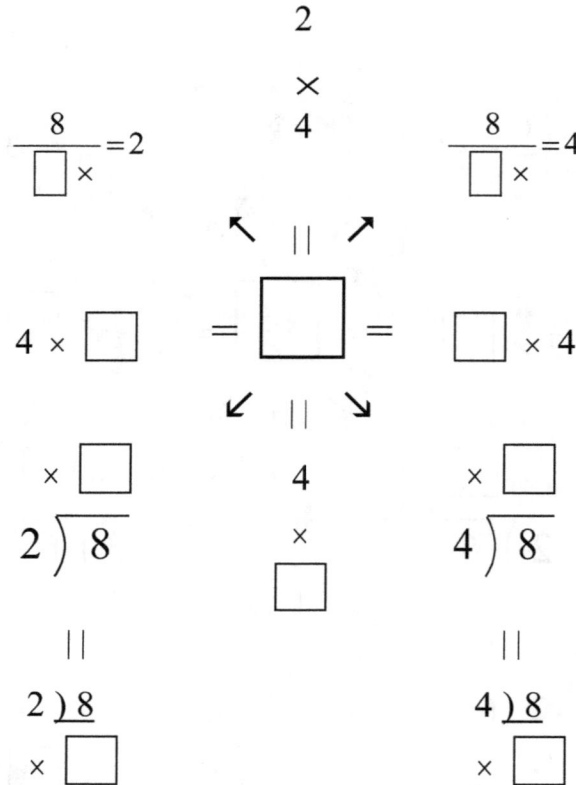

$$2$$
$$\times$$
$$4$$

$$\frac{8}{\Box \times} = 2$$

$$\frac{8}{\Box \times} = 4$$

$$\nwarrow \quad || \quad \nearrow$$

$$4 \times \Box \quad = \boxed{} = \quad \Box \times 4$$

$$\swarrow \quad || \quad \searrow$$

$$\times \Box$$

$$2\overline{)\,8}$$

$$4$$
$$\times$$
$$\Box$$

$$\times \Box$$

$$4\overline{)\,8}$$

$$||$$ $$||$$

$$2\,)\,\underline{8}$$ $$4\,)\,\underline{8}$$

$$\times \Box$$ $$\times \Box$$

Ho Math Chess 何数棋谜 妈!我会棋谜式乘法啦!
Mom! I Learn Multiplication Using Math-Chess-Puzzles Connection!

Student's Name _____ Date _____

2007 - 2018 © Frank Ho, Amanda Ho, All rights reserved.　　www.homathchess.com

Different ways of writing multiplication (Learning division while doing multiplications)

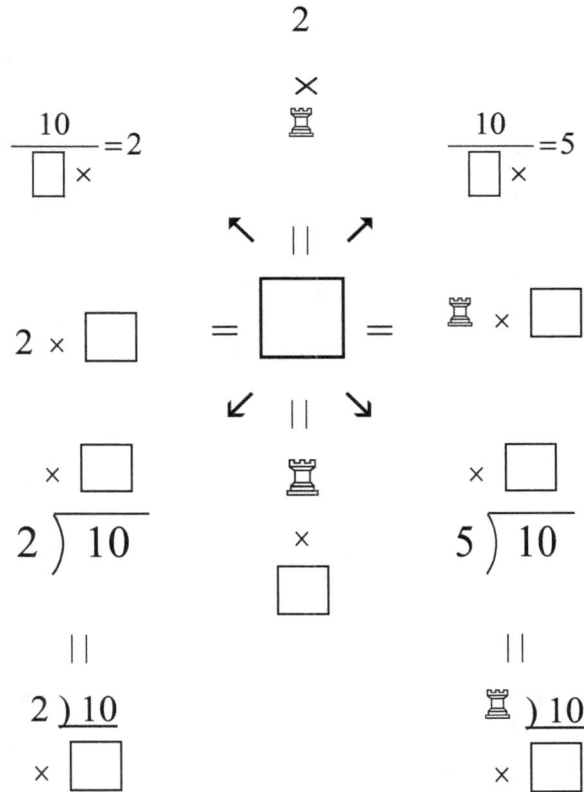

2007 - 2018 © Frank Ho, Amanda Ho, All rights reserved.　　www.homathchess.com

Different ways of writing multiplication (Learning division while doing multiplications)

$$\frac{12}{\boxed{}\times}=2 \qquad \begin{array}{c}2\\ \times\\ 6\end{array} \qquad \frac{12}{\boxed{}\times}=6$$

$$2 \times \boxed{} \qquad = \quad \boxed{} \quad = \qquad 6 \times \boxed{}$$

$$\begin{array}{r} \times\ \boxed{}\\ 2\,\overline{)\,12}\end{array} \qquad \begin{array}{c}6\\ \times\\ \boxed{}\end{array} \qquad \begin{array}{r} \times\ \boxed{}\\ 6\,\overline{)\,12}\end{array}$$

$$\begin{array}{c} 2\,\underline{)\,12}\\ \times\ \boxed{}\end{array} \qquad\qquad\qquad \begin{array}{c} 6\,\underline{)\,12}\\ \times\ \boxed{}\end{array}$$

2007 - 2018 © Frank Ho, Amanda Ho, All rights reserved.   www.homathchess.com

Different ways of writing multiplication (Learning division while doing multiplications)

2007 - 2018 © Frank Ho, Amanda Ho, All rights reserved.   www.homathchess.com

Different ways of writing multiplication (Learning division while doing multiplications)

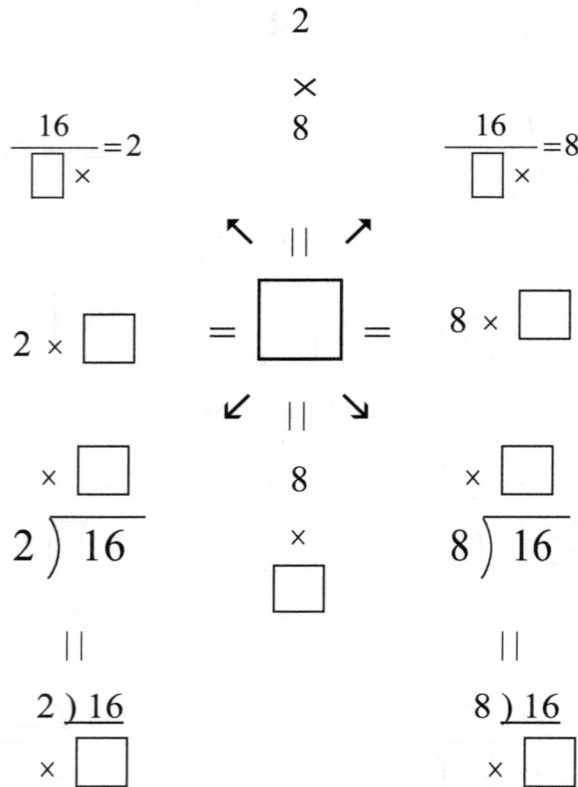

$$2 \times 8$$

$$\frac{16}{\square \times} = 2 \qquad \frac{16}{\square \times} = 8$$

$$2 \times \square \quad = \boxed{} = \quad 8 \times \square$$

$$\times \square$$
$$2\overline{\smash{)}\,16}$$

$$8 \times \square$$

$$\times \square$$
$$8\overline{\smash{)}\,16}$$

$$\parallel \qquad\qquad\qquad \parallel$$

$$2\,\underline{)\,16} \qquad\qquad 8\,\underline{)\,16}$$
$$\times \square \qquad\qquad\qquad \times \square$$

Student's Name _____ Date _____

2007 - 2018 © Frank Ho, Amanda Ho, All rights reserved.   www.homathchess.com

Different ways of writing multiplication (Learning division while doing multiplications)

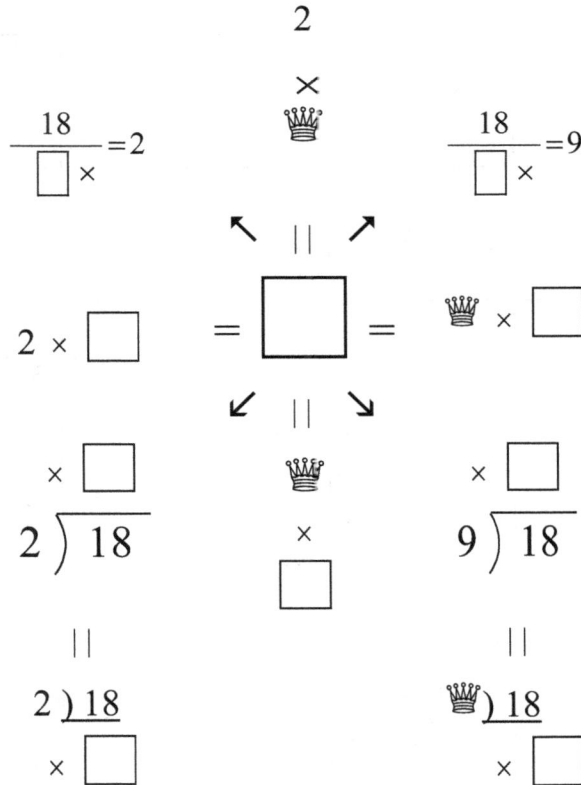

2007 - 2018 © Frank Ho, Amanda Ho, All rights reserved.　www.homathchess.com

2 times

Fill in each □ with >, <, or =.

$8 \times 2 \ \square\ 7 \times 2$	$8 \times 2 \ \square\ 9 \times 2$	$8 \times 2 \ \square\ 7 \times 2$
$1 \times 2 \ \square\ 2 \times 1$	$4 \times 2 \ \square\ 2 \times 5$	$2 \times 2 \ \square\ 2 \times 3$
$2 \times 7 \ \square\ 7 \times 2$	$5 \times 2 \ \square\ 2 \times 6$	$5 \times 2 \ \square\ 4 \times 2$
$8 \times 2 \ \square\ 7 \times 2$	$9 \times 2 \ \square\ 9 \times 2$	$3 \times 2 \ \square\ 2 \times 2$
$8 \times 2 \ \square\ 9 \times 2$	$5 \times 2 \ \square\ 2 \times 5$	$6 \times 2 \ \square\ 7 \times 2$
$8 \times 2 \ \square\ 6 \times 2$	$6 \times 2 \ \square\ 4 \times 2$	$6 \times 2 \ \square\ 2 \times 6$
$8 \times 2 \ \square\ 7 \times 2$	$9 \times 2 \ \square\ 2 \times 9$	$2 \times 2 \ \square\ 7 \times 2$
$3 \times 2 \ \square\ 2 \times 3$	$8 \times 2 \ \square\ 2 \times 5$	$2 \times 2 \ \square\ 2 \times 3$
$9 \times 2 \ \square\ 7 \times 2$	$4 \times 2 \ \square\ 2 \times 3$	$8 \times 2 \ \square\ 2 \times 4$
$8 \times 2 \ \square\ 7 \times 2$	$5 \times 2 \ \square\ 7 \times 2$	$6 \times 2 \ \square\ 5 \times 2$

2007 - 2018 © Frank Ho, Amanda Ho, All rights reserved.   www.homathchess.com

Counting 3's multiples (Tripling)

Circle the following 3's multiples.

1 2 3 4 5 6 7 8 9 10 11 12 13 14 15 16 17 18
19 20 21 22 23 24 25 26 27

Fill in the following ☐ with a number.

Sequence	1	2	3	4	5	6	7	8	9
Add 3	☐	6	9	☐	15	18	☐	☐	27

Sequence	1	2	♗	4	♖	6	7	8	♕
Add 3	3	6	☐	☐	15	18	☐	☐	27

Sequence	1	2	♗	4	♖	6	7	8	♕
Add 3	☐	6	♕	☐	15	☐	☐	☐	27

Sequence	1	2	3	4	5	6	7	8	9
Add 3	☐	☐	9	☐	15	18	☐	☐	27

Sequence	1	2	♗	4	♖	6	7	8	♕
Add 3	☐	6	☐	☐	☐	18	☐	☐	27

Mom! I Learn Multiplication Using Math-Chess-Puzzles Connection!

Student's Name _____ Date _____

2007 - 2018 © Frank Ho, Amanda Ho, All rights reserved.　www.homathchess.com

3 times

2+2+2 = □	= 3 × 2 =	2 × 3 = □		2 2 + 2 □ = 3 × 2 = □
3+3+3 = □	= 3 × 3 =	♗ × ♗ = □		♗ ♗ + ♗ □ = 3 × 3 = □
4+4+4 = □	= 3 × 4 =	4 × 3 = □		4 4 + 4 □ = 3 × 4 = □
5+5+5 = □	= 3 × ♖ =	5 × 3 = □		♖ ♖ + ♖ □ = 3 × ♖ = □

Ho Math Chess 何数棋谜 妈！我会棋谜式乘法啦！
Mom! I Learn Multiplication Using Math-Chess-Puzzles Connection!

Student's Name _____ Date _____

2007 - 2018 © Frank Ho, Amanda Ho, All rights reserved.   www.homathchess.com

3 times

$6+6+6 =$ $\square$	$= 3 \times 6 =$	$6 \times 3 =$ $\square$	$\begin{array}{r} 6 \\ 6 \\ +\,6 \\ \hline \end{array}$ $\square = \text{♗} \times 6 = \square$	
$7+7+7 =$ $\square$	$= 3 \times 7 =$	$7 \times 3 =$ $\square$	$\begin{array}{r} 7 \\ 7 \\ +\,7 \\ \hline \end{array}$ $\square = \text{♗} \times 7 = \square$	
$8+8+8 =$ $\square$	$= 3 \times 8 =$	$8 \times 3 =$ $\square$	$\begin{array}{r} 8 \\ 8 \\ +\,8 \\ \hline \end{array}$ $\square = 3 \times 8 = \square$	
$9+9+9 =$ $\square$	$= 3 \times 9 =$	$9 \times 3 =$ $\square$	$\begin{array}{r} 9 \\ 9 \\ +\,9 \\ \hline \end{array}$ $\square = 3 \times \text{♕} = \square$	

2007 - 2018 © Frank Ho, Amanda Ho, All rights reserved.　　www.homathchess.com

3 times

$3 \times 1 = \square$	Three times one is $\square$	$1 \times 3 = \square$	One times three is $\square$
$3 \times 2 = \square$	Three times two is $\square$	$2 \times 3 = \square$	Two times three is $\square$
$3 \times 3 = \square$	Three times three is $\square$	$3 \times 3 = \square$	Three times three is $\square$
$3 \times 4 = \square$	Three times four is $\square$	$4 \times 3 = \square$	Four times three is $\square$
$3 \times 5 = \square$	Three times five is $\square$	$5 \times 3 = \square$	Five times three is $\square$
$3 \times 6 = \square$	Three times six is $\square$	$6 \times 3 = \square$	Six times three is $\square$
$3 \times 7 = \square$	**Three times seven is** $\square$	$7 \times 3 = \square$	Seven times three is $\square$
$3 \times 8 = \square$	Three times eight is $\square$	$8 \times 3 = \square$	Eight times three is $\square$
$3 \times 9 = \square$	Three times nine is $\square$	♛ $\times 3 = \square$	Nine times three is $\square$

$$\begin{array}{ccccc}
3 & 1 & 2 & 3 & 3 \\
\underline{\times\,1} & \underline{\times\,3} & \underline{\times\,3} & \underline{\times\,2} & \underline{\times\,3} \\
\square & \square & \square & \square & \square
\end{array}$$

$$\begin{array}{ccccc}
5 & 3 & 7 & 3 & ♛ \\
\underline{\times\,3} & \underline{\times\,6} & \underline{\times\,3} & \underline{\times\,8} & \underline{\times\,3} \\
\square\square & \square\square & \square\square & \square\square & \square\square
\end{array}$$

$$\begin{array}{ccccc}
4 & 3 & 5 & 3 & 3 \\
\underline{\times\,3} & \underline{\times\,8} & \underline{\times\,3} & \underline{\times\,6} & \underline{\times\,9} \\
\square\square & \square\square & \square\square & \square\square & \square\square
\end{array}$$

Mom! I Learn Multiplication Using Math-Chess-Puzzles Connection!

Student's Name _____　Date _____

2007 - 2018 © Frank Ho, Amanda Ho, All rights reserved.　www.homathchess.com

3	1	4	3	3
X 1	X 3	X ♗	X 4	X 3
☐	☐	☐☐	☐☐	☐

♖	3	♘	8	♗
X 3	X 6	X 7	X 3	X 9
☐☐	☐☐	☐☐	☐☐	☐☐

4	3	5	6	7
X ♘	X 4	X 3	X ♗	X ♘
☐☐	☐☐	☐☐	☐☐	☐☐

8	♕	3	♘	3
X 3	X 3	X 4	X 8	X 5
☐☐	☐☐	☐☐	☐☐	☐☐

4	3	3	3	3
X 3	X 6	X 7	X 3	X 9
☐☐	☐☐	☐☐	☐	☐☐

Ho Math Chess 何数棋谜 妈!我会棋谜式乘法啦!

Mom! I Learn Multiplication Using Math-Chess-Puzzles Connection!

Student's Name _____ Date _____

2007 - 2018 © Frank Ho, Amanda Ho, All rights reserved. www.homathchess.com

Oral practice

Three one three	3 1 ☐	1 1 × 3 ☐☐
Three two six (is half-dozen)	3 2 ☐	2 2 × 3 ☐☐
Three three nine	3 3 ☐	3 3 × ♘ ☐☐
Three four twelve (is a dozen)	3 4 ☐	¹ 4 4 × 3 ☐☐☐
Three five fifteen	3 5 ☐	¹ 5 5 × ♘ ☐☐☐

Student's Name _____ Date _____

2007 - 2018 © Frank Ho, Amanda Ho, All rights reserved.   www.homathchess.com

Oral practice

Three six eighteen	♗ 6 ☐	1 6 6 × 3 ☐ ☐ ☐
Three seven twenty-one	3 7 ☐	7 7 × ♗ ☐ ☐ ☐
Three eight twenty-four	3 8 ☐	8 8 × 3 ☐ ☐ ☐
Three nine twenty-seven	♘ 9 ☐	9 9 × 3 ☐ ☐ ☐
Three five fifteen	3 5 ☐	5 5 × 3 ☐ ☐ ☐
Three six eighteen	3 6 ☐	6 6 × ♘ ☐ ☐ ☐

2007 - 2018 © Frank Ho, Amanda Ho, All rights reserved.   www.homathchess.com

Preparing for division

□	□	□	□	□
X 2	X ♞	X 4	X 2	X ♝
6	♛	1 2	6	1 2

□	□	□	□	□
X 4	X 6	X 3	X 3	X 2
2 8	1 8	2 1	1 5	6

□	□	□	□	□
X 9	X 3	X ♝	X ♞	X 7
2 7	1 2	2 1	1 5	2 1

□	□	□	□	□
X ♜	X 3	X 4	X ♞	X 2
1 5	6	1 2	♛	6

□	□	□	□	□
X 4	X 5	X ♝	X 2	X 3
1 2	1 5	1 8	1 2	2 1

Ho Math Chess 何数棋谜 妈！我会棋谜式乘法啦！
Mom! I Learn Multiplication Using Math-Chess-Puzzles Connection!

Student's Name _____ Date _____

2007 - 2018 © Frank Ho, Amanda Ho, All rights reserved.   www.homathchess.com

Preparing for division

□ X ♘ = 3	X □ 3)3	□)3 X ♘
□ X 3 = 6	X □ 3)6	□)6 X 3
□ X ♗ = 9	X □ 3)9	□)♕ X ♕
□ X 3 = 12	X □ 3)12	□)12 X 3
□ X ♘ = 15	X □ 3)15	□)15 X ♗
□ X ♗ = 18	X □ 3)18	□)18 X ♘
□ X ♘ = 21	X □ 3)21	□)21 X ♘

Ho Math Chess   何数棋谜   妈！我会棋谜式乘法啦！
Mom! I Learn Multiplication Using Math-Chess-Puzzles Connection!

Student's Name _____ Date _____

2007 - 2018 © Frank Ho, Amanda Ho, All rights reserved.   www.homathchess.com

Fill in ☐ with answer.

Times	Grouping	Addition
$3 \times \$1 = $ ☐	3 of ☐ $= ♗$	$\$1 + \$1 + \$1 = $ ☐
$1 \times \$3 = $ ☐	1 of ☐ $= 3$	3 of $\$1 = $ ☐

Fill in ☐ with answer.

Expression	Grouping	Addition
$3 \times \$2$	3 of ☐ $= 6$	$\$2 + \$2 + \$2 = $ ☐
$2 \times \$3$	2 of ☐ $= 6$	$\$3 + \$3 = $ ☐

$3 \times 1 = $ ☐ $= 1 \times 3 = $ ☐	$1 \times 3 = $ ☐ $= 3 \times 1 = $ ☐
$3 \times $ ☐ $= 6 = 2 \times $ ☐ $= $ ☐	$2 \times $ ☐ $= 6 = 3 \times $ ☐ $= $ ☐

$3\ 1$ ☐	$3\ 5$ ☐	$3\ 9$ ☐	$♗\ 4$ ☐	$3\ 8$ ☐
$3\ 2$ ☐	$♗\ 6$ ☐	$♗\ 1$ ☐	$♗\ ♖$ ☐	$♘\ 9$ ☐
$♘\ ♗$ ☐	$3\ 7$ ☐	$3\ 2$ ☐	$3\ 6$ ☐	$3\ 1$ ☐
$3\ 4$ ☐	$3\ 8$ ☐	$3\ ♘$ ☐	$3\ 7$ ☐	$3\ 2$ ☐

2007 - 2018 © Frank Ho, Amanda Ho, All rights reserved.   www.homathchess.com

Fill in ☐ with answer.

Times	Grouping	Addition
3 × $3 = ☐	3 of ☐ = 9	$3 + $3 + $3 = ☐
3 × $3 = ☐	3 of ☐ = 9	$3 + $3 + $♗ = ☐

Fill in ☐ with answer.

Expression	Grouping	Addition
3 × $4	♗ of ☐ = 12	$4 + $4 + $4 = ☐
4 × $3	4 of ☐ = 12	$3 + $3 + $3 + $♘ = ☐

♘ × 3 = ☐ = 3 × 3 = ☐	♗ × 3 = ☐ = 3 × 3 = ☐
3 × ☐ = 12 = 4 × ☐ = ☐	4 × ☐ = 12 = 3 × ☐ = ☐

3 1 ☐	3 5 ☐	♗ 9 ☐	3 4 ☐	3 8 ☐
3 2 ☐	3 6 ☐	3 1 ☐	♘ 5 ☐	3 9 ☐
3 3 ☐	♘ 7 ☐	3 2 ☐	3 6 ☐	3 1 ☐
3 4 ☐	3 8 ☐	♗ ♗ ☐	3 7 ☐	♘ 2 ☐

Student's Name _____ Date _____

2007 - 2018 © Frank Ho, Amanda Ho, All rights reserved.   www.homathchess.com

Fill in ☐ with answer.

Times	Grouping	Addition
$3 \times \$5 = \square$	3 of $\square$ = 15	$\$5 + \$♖ + \$5 = \square$
$5 \times \$♘ = \square$	5 of $\square$ = 15	$\$3 + \$3 + \$♗ + \$3 + \$3 = \square$

Times	Grouping	Addition
$3 \times \$6 = \square$	3 of $\square$ = 18	$\$6 + \$6 + \$6 = \square$
$6 \times \$3 = \square$	6 of $\square$ = 18	$\$3 + \$3 + \$3 + \$3 + \$3 + \$3 = \square$

$♘ \times 5 = \square = 5 \times 3 = \square$	$♖ \times 3 = \square = ♘ \times 5 = \square$
$3 \times \square = 18 = 6 \times \square = \square$	$6 \times \square = 18 = ♗ \times \square = \square$

♘ 8 $\square$	3 5 $\square$	3 9 $\square$	3 4 $\square$	♗ 8 $\square$
3 2 $\square$	3 6 $\square$	♘ 1 $\square$	3 ♖ $\square$	3 9 $\square$
3 3 $\square$	♘ 7 $\square$	3 2 $\square$	3 6 $\square$	♘ 1 $\square$
♗ 4 $\square$	3 8 $\square$	3 3 $\square$	♘ 7 $\square$	3 2 $\square$

2007 - 2018 © Frank Ho, Amanda Ho, All rights reserved.   www.homathchess.com

Fill in ☐ with answer.

Times	Grouping	Addition
3 × \$7 = ☐	3 of ☐ = 21	\$7 + \$7 + \$7 = ☐
7 × \$3 = ☐	7 of ☐ = 21	\$3 + \$3 + \$3 + \$3 + \$3 + \$3 + \$3 = ☐

Times	Grouping	Addition
3 × \$8 = ☐	3 of ☐ = 24	\$8 + \$8 + \$8 = ☐
8 × \$3 = ☐	8 of ☐ = 24	\$3 + \$3 + \$3 + \$3 + \$3 + \$3 + \$3 + \$3 = ☐

3 × 7 = ☐ = 7 × ♗ = ☐	7 × 3 = ☐ = 3 × 7 = ☐
3 × ☐ = 24 = 8 × ☐ = ☐	8 × ☐ = 24 = ♗ × ☐ = ☐

3 1 ☐	3 5 ☐	3 9 ☐	3 4 ☐	3 8 ☐
3 2 ☐	♗ 6 ☐	3 ♙ ☐	♗ 5 ☐	3 9 ☐
3 ♗ ☐	3 7 ☐	3 2 ☐	3 6 ☐	3 ♙ ☐
3 4 ☐	3 8 ☐	♗ ♗ ☐	3 7 ☐	3 2 ☐

Ho Math Chess   何数棋谜   妈！我会棋谜式乘法啦！
Mom! I Learn Multiplication Using Math-Chess-Puzzles Connection!

Student's Name _____ Date _____

2007 - 2018 © Frank Ho, Amanda Ho, All rights reserved.   www.homathchess.com

Fill in ☐ with answer.

Times	Grouping	Addition
3 × \$9 = ☐	3 of ☐ = 27	\$9 + \$9 + \$9 = ☐
9 × \$3 = ☐	9 of ☐ = 27	\$3 + \$3 + \$3 + \$3 + \$3 + \$3 + \$3 + \$3 + \$3 = ☐

Times	Grouping	Addition
3 × \$8 = ☐	3 of ☐ = 24	\$8 + \$8 + \$8 = ☐
8 × \$3 = ☐	8 of ☐ = 24	\$3 + \$3 + \$3 + \$3 + \$3 + \$3 + \$3 + \$3 = ☐

3 × 8 = ☐ = 8 × 3 = ☐	8 × 3 = ☐ = 3 × 8 = ☐
3 × ☐ = 27 = 9 × ☐ = ☐	♛ × ☐ = 27 = 3 × ☐ = ☐

3 1 ☐	3 5 ☐	3 ♛ ☐	3 4 ☐	♞ 8 ☐
3 2 ☐	♗ 6 ☐	3 1 ☐	♗ 5 ☐	3 9 ☐
3 ♗ ☐	3 7 ☐	♗ 2 ☐	3 6 ☐	3 ♙ ☐
3 4 ☐	3 8 ☐	3 3 ☐	♗ 7 ☐	3 2 ☐

Mom! I Learn Multiplication Using Math-Chess-Puzzles Connection!

Student's Name _____ Date _____

2007 - 2018 © Frank Ho, Amanda Ho, All rights reserved.   www.homathchess.com

Preparing for division

☐ X ♝ = 24	X ☐ 3)24	☐)24 X 3
☐ X 3 = 27	X ☐ 3)27	☐)27 X 3
☐ X ♝ = ♛	X ☐ 3)9	☐)♛ X 3
☐ X 3 = 12	X ☐ 3)12	☐)12 X 3
☐ X ♝ = 15	X ☐ 3)15	☐)15 X ♞
☐ X 3 = 18	X ☐ 3)18	☐)18 X 3
☐ X 3 = 21	X ☐ 3)21	☐)21 X ♞

2007 - 2018 © Frank Ho, Amanda Ho, All rights reserved.   www.homathchess.com

Cross multiplication

12 ⬈⬉ 12 $\frac{6}{2} = \frac{6}{2}$	□ ⬈⬉ □ $\frac{3}{3} = \frac{2}{2}$	□ ⬈⬉ □ $\frac{3}{3} = \frac{3}{3}$	□ ⬈⬉ □ $\frac{3}{3} = \frac{4}{4}$
□ ⬈⬉ □ $\frac{3}{3} = \frac{5}{5}$	□ ⬈⬉ □ $\frac{3}{3} = \frac{6}{6}$	□ ⬈⬉ □ $\frac{3}{3} = \frac{7}{7}$	□ ⬈⬉ □ $\frac{3}{3} = \frac{9}{9}$
□ ⬈⬉ □ $\frac{3}{3} = \frac{5}{5}$	□ ⬈⬉ □ $\frac{3}{3} = \frac{4}{4}$	□ ⬈⬉ □ $\frac{3}{3} = \frac{7}{7}$	□ ⬈⬉ □ $\frac{3}{3} = \frac{8}{8}$
□ ⬈⬉ □ $\frac{3}{3} = \frac{6}{6}$	□ ⬈⬉ □ $\frac{3}{3} = \frac{8}{8}$	□ ⬈⬉ □ $\frac{3}{3} = \frac{9}{9}$	□ ⬈⬉ □ $\frac{3}{3} = \frac{3}{3}$

2007 - 2018 © Frank Ho, Amanda Ho, All rights reserved.   www.homathchess.com

Different ways of writing multiplication (Learning division while doing multiplications)

$$\frac{6}{\square \times} = 2 \qquad \text{♘} \times 2 \qquad \frac{6}{\square \times} = 3$$

$$2 \times \square \quad = \boxed{} = \quad \text{♗} \times \square$$

$$\times \square \qquad 2 \qquad \times \square$$

$$2\overline{)6} \qquad \times \square \qquad 3\overline{)6}$$

$$2\,)6 \qquad\qquad \text{♘}\,)6$$

$$\times \square \qquad\qquad\qquad \times \square$$

2007 - 2018 © Frank Ho, Amanda Ho, All rights reserved.   www.homathchess.com

Different ways of writing multiplication (Learning division while doing multiplications)

2007 - 2018 © Frank Ho, Amanda Ho, All rights reserved.   www.homathchess.com

Different ways of writing multiplication (Learning division while doing multiplications)

$$\frac{12}{\boxed{}\times} = 3 \qquad \overset{\times}{4} \qquad \frac{12}{\boxed{}\times} = 4$$

$$\nwarrow \;\; || \;\; \nearrow$$

$$3 \times \boxed{} \;\; = \;\; \boxed{} \;\; = \;\; \boxed{} \times 4$$

$$\swarrow \;\; || \;\; \searrow$$

$$\times \boxed{} \qquad\qquad 4 \qquad\qquad \times \boxed{}$$

$$3 \overline{)\,12} \qquad\qquad \overset{\times}{\underset{\boxed{}}{}} \qquad\qquad 4 \overline{)\,12}$$

$$|| \qquad\qquad\qquad\qquad\qquad ||$$

$$\overline{)\,12} \qquad\qquad\qquad\qquad 4\,\overline{)\,12}$$

$$\times \boxed{} \qquad\qquad\qquad\qquad \times \boxed{}$$

2007 - 2018 © Frank Ho, Amanda Ho, All rights reserved.　　www.homathchess.com

Different ways of writing multiplication (Learning division while doing multiplications)

$$\frac{15}{\square\times}=3 \qquad\qquad \frac{15}{\square\times}=5$$

$$3\times\square \qquad = \quad \boxed{\ } \quad = \qquad 5\times\square$$

$$3\overline{\smash{\big)}\,15} \qquad\qquad 5\overline{\smash{\big)}\,15}$$

2007 - 2018 © Frank Ho, Amanda Ho, All rights reserved.   www.homathchess.com

Different ways of writing multiplication (Learning division while doing multiplications)

2007 - 2018 © Frank Ho, Amanda Ho, All rights reserved.   www.homathchess.com

Different ways of writing multiplication (Learning division while doing multiplications)

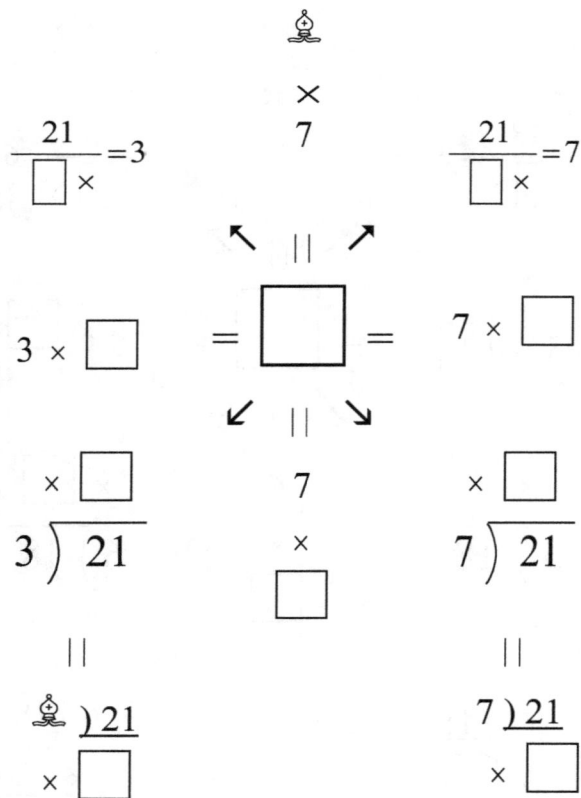

$$\frac{21}{\square \times} = 3 \qquad\qquad \frac{21}{\square \times} = 7$$

$$\unicode{x2659} \times 7$$

$$3 \times \square \quad = \boxed{} = \quad 7 \times \square$$

$$\begin{array}{c} \times \square \\ 3\,\overline{)\,21} \end{array} \qquad 7 \times \square \qquad \begin{array}{c} \times \square \\ 7\,\overline{)\,21} \end{array}$$

$$\unicode{x2659}\,\overline{)\,21} \qquad\qquad 7\,\overline{)\,21}$$
$$\times \square \qquad\qquad\qquad \times \square$$

2007 - 2018 © Frank Ho, Amanda Ho, All rights reserved.　www.homathchess.com

Different ways of writing multiplication (Learning division while doing multiplications)

$$\frac{24}{\Box\times}=3 \qquad \knight\times 8 \qquad \frac{24}{\Box\times}=8$$

$$\knight\times\Box = \boxed{} = 8\times\Box$$

$$\times\Box$$
$$3\overline{)24} \qquad 8\times\Box \qquad 8\overline{)24}$$

$$\knight\overline{)24} \qquad\qquad 8\overline{)24}$$
$$\times\Box \qquad\qquad \times\Box$$

2007 - 2018 © Frank Ho, Amanda Ho, All rights reserved.   www.homathchess.com

Different ways of writing multiplication (Learning division while doing multiplications)

$$\frac{27}{\boxed{}\times} = 3 \qquad \frac{27}{\boxed{}\times} = 9$$

$$\text{♗} \times \boxed{} = \boxed{} = 9 \times \boxed{}$$

$$3\overline{)27} \qquad 9 \times \boxed{} \qquad 9\overline{)27}$$

$$3\underline{)27} \qquad \text{♛}\underline{)27}$$
$$\times \boxed{} \qquad \times \boxed{}$$

Ho Math Chess　何数棋谜　妈！我会棋谜式乘法啦！

Mom! I Learn Multiplication Using Math-Chess-Puzzles Connection!

Student's Name _____ Date _____

2007 - 2018 © Frank Ho, Amanda Ho, All rights reserved.　　www.homathchess.com

3 times

Fill in each □ with >, <, or =.

8×3 □ 7×3	6×3 □ 5×3	8×3 □ 9×3
1×3 □ ♞ $\times 1$	9×3 □ $3 \times$ ♛	4×3 □ 3×5
3×7 □ $7 \times$ ♝	5×3 □ $4 \times$ ♞	♜ $\times 3$ □ 3×6
8×3 □ 7×3	8×3 □ $7 \times$ ♝	9×3 □ 9×3
$8 \times$ ♝ □ 9×3	3×2 □ 2×3	5×3 □ ♞ $\times$ ♜
8×3 □ 6×3	$6 \times$ ♝ □ 3×6	6×3 □ 4×3
8×3 □ $7 \times$ ♝	2×3 □ $7 \times$ ♞	$9 \times$ ♞ □ 3×9
3×3 □ 2×3	3×3 □ 3×3	8×3 □ $3 \times$ ♜
$9 \times$ ♞ □ 7×3	8×3 □ 3×4	$4 \times$ ♝ □ 3×3
8×3 □ 7×3	6×3 □ ♜ $\times$ ♞	5×3 □ $7 \times$ ♝

Mom! I Learn Multiplication Using Math-Chess-Puzzles Connection!

Student's Name _____ Date _____

2007 - 2018 © Frank Ho, Amanda Ho, All rights reserved.   www.homathchess.com

Counting 4's multiples

Write 4's multiples.

4, 8, 12, □,□,□,□,□,□

Fill in the following each □ with a number.

Sequence	1	2	3	4	♜	6	7	8	♛
Add 4	□	8	□	16	□	24	□	32	□

Sequence	♟	2	♝	4	5	6	7	8	9
Add 4	4	□	12	□	20	□	28	□	36

Sequence	1	2	3	4	♜	6	7	8	9
Add 4	□	8	□	16	□	24	□	32	□

Sequence	♟	2	♝	4	5	6	7	8	♛
Add 4	4	□	12	□	20	□	28	□	36

Sequence	1	2	3	4	♜	6	7	8	9
Add 4	□	□	□	16	□	24	□	32	□

Sequence	♟	2	♝	4	5	6	7	8	9
Add 4	□	8	□	□	□	24	□	32	□

2007 - 2018 © Frank Ho, Amanda Ho, All rights reserved.　www.homathchess.com

4 times

4 × 1 = ☐	Four times one is ☐	1 × 3 = ☐	One times four is ☐
4 × 2 = ☐	Four times two is ☐	2 × 3 = ☐	Two times four is ☐
4 × 3 = ☐	Four times three is ☐	♗ × 3 = ☐	Three times four is ☐
4 × 4 = ☐	Four times four is ☐	4 × 3 = ☐	Four times four is ☐
4 × ♖ = ☐	Four times five is ☐	♖ × 3 = ☐	Five times four is ☐
4 × 6 = ☐	Four times six is ☐	6 × 3 = ☐	Six times four is ☐
4 × 7 = ☐	Four times seven is ☐	7 × 3 = ☐	Seven times four is ☐
4 × 8 = ☐	Four times eight is ☐	8 × 3 = ☐	Eight times four is ☐
4 × 9 = ☐	Four times nine is ☐	9 × ♗ = ☐	Nine times four is ☐

Row 1:

$$4 \times 1 = \square \qquad ♙ \times 4 = \square \qquad 2 \times 4 = \square \qquad 4 \times 2 = \square \qquad 4 \times ♗ = \square$$

Row 2:

$$4 \times 3 = \square \qquad 4 \times 6 = \square \qquad 7 \times 4 = \square \qquad 4 \times 8 = \square \qquad ♕ \times 4 = \square$$

Row 3:

$$4 \times ♖ = \square \qquad 4 \times 8 = \square \qquad ♖ \times 4 = \square \qquad 4 \times 6 = \square \qquad 4 \times ♕ = \square$$

Ho Math Chess　　何数棋谜　妈!我会棋谜式乘法啦!

Mom! I Learn Multiplication Using Math-Chess-Puzzles Connection!

Student's Name _____ Date _____

2007 - 2018 © Frank Ho, Amanda Ho, All rights reserved.　　www.homathchess.com

4	1	4	2	4
X ♙	X 4	X 2	X 4	X ♗

5	4	4	8	4
X 4	X 6	X 7	X 4	X 9

4	♖	8	4	♕
X 6	X 4	X 4	X 6	X 4

5	4	7	4	9
X 4	X 6	X 4	X 8	X 4

4	4	4	4	4
X ♖	X 8	X 2	X 6	X ♕

66

2007 - 2018 © Frank Ho, Amanda Ho, All rights reserved.   www.homathchess.com

Oral practice

four one four	4 1 □	$\begin{array}{r} 1\,1 \\ \times\quad 4 \\ \hline \square\square \end{array}$
four two eight	4 2 □	$\begin{array}{r} 2\,2 \\ \times\quad 4 \\ \hline \square\square \end{array}$
four three twelve (is a dozen)	4 3 □	$\begin{array}{r} 3\,3 \\ \times\quad 4 \\ \hline \square\square\square \end{array}$
four four sixteen	4 4 □	$\begin{array}{r} ^1\;\;\, \\ 4\,4 \\ \times\quad 4 \\ \hline \square\square\square \end{array}$
four five twenty	4 ♖ □	$\begin{array}{r} ^2\;\;\, \\ 5\,5 \\ \times\quad 4 \\ \hline \square\square\square \end{array}$

2007 - 2018 © Frank Ho, Amanda Ho, All rights reserved.　www.homathchess.com

Oral practice

four six twenty-four	4 6 ☐	$\begin{array}{r} 2 \\ 6\,6 \\ \times\quad 4 \\ \hline \square\square\square \end{array}$
four seven twenty-eight	4 7 ☐	$\begin{array}{r} 7\,7 \\ \times\quad 4 \\ \hline \square\square\square \end{array}$
four eight thirty-two	4 8 ☐	$\begin{array}{r} 8\,8 \\ \times\quad 4 \\ \hline \square\square\square \end{array}$
four nine thirty-six	4 ♛ ☐	$\begin{array}{r} 9\,9 \\ \times\quad 4 \\ \hline \square\square\square \end{array}$
four five twenty	4 ♜ ☐	$\begin{array}{r} 5\,5 \\ \times\quad 4 \\ \hline \square\square\square \end{array}$
four six twenty-four	4 6 ☐	$\begin{array}{r} 6\,6 \\ \times\quad 4 \\ \hline \square\square\square \end{array}$

Student's Name _____ Date _____

2007 - 2018 © Frank Ho, Amanda Ho, All rights reserved.   www.homathchess.com

Fill in ☐ with answer.

Times	Grouping	Addition
$4 \times \$1 = \square$	4 of $\square = 4$	$\$1 + \$1 + \$1 + \$1 = \square$
$1 \times \$4 = \square$	1 of $\square = 4$	4 of $\$1 = \square$

Fill in ☐ with answer.

Expression	Grouping	Addition
$4 \times \$2$	4 of $\square = 8$	$\$2 + \$2 + \$2 + \$2 = \square$
$2 \times \$4$	2 of $\square = 8$	$\$4 + \$4 = \square$

$4 \times 1 = \square = 1 \times 4 = \square$	$1 \times 4 = \square = 4 \times 1 = \square$
$4 \times \square = 8 = 2 \times \square = \square$	$2 \times \square = 8 = 4 \times \square = \square$

4 1 ☐	4 5 ☐	4 9 ☐	4 4 ☐	4 8 ☐
4 2 ☐	4 6 ☐	4 1 ☐	4 5 ☐	4 ♛ ☐
4 3 ☐	4 7 ☐	4 2 ☐	4 6 ☐	4 1 ☐
4 4 ☐	4 8 ☐	4 3 ☐	4 7 ☐	4 2 ☐

Ho Math Chess 何数棋谜 妈！我会棋谜式乘法啦！

Mom! I Learn Multiplication Using Math-Chess-Puzzles Connection!

Student's Name _____ Date _____

2007 - 2018 © Frank Ho, Amanda Ho, All rights reserved. www.homathchess.com

Fill in ☐ with answer.

Times	Grouping	Addition
4 × $3 = ☐	4 of ☐ = 12	$3 + $3 + $3 + $3 = ☐
3 × $4 = ☐	3 of ☐ = 12	$4 + $4 + $4 = ☐

Fill in ☐ with answer.

Expression	Grouping	Addition
4 × $4	4 of ☐ = 16	$4 + $4 + $4 + $4 = ☐
4 × $4	4 of ☐ = 16	$4 + $4 + $4 + $4 = ☐

3 × 4 = ☐ = 4 × 3 = ☐	4 × 3 = ☐ = 3 × 4 = ☐
4 × ☐ = 16 = 4 × ☐ = ☐	4 × ☐ = 16 = 4 × ☐ = ☐

4 ♙ ☐	4 5 ☐	4 9 ☐	4 4 ☐	4 8 ☐
4 2 ☐	4 6 ☐	4 ♙ ☐	4 5 ☐	4 ♛ ☐
4 ♞ ☐	4 7 ☐	4 2 ☐	4 6 ☐	4 1 ☐
4 4 ☐	4 8 ☐	4 ♞ ☐	4 7 ☐	4 2 ☐

Student's Name _____ Date _____

2007 - 2018 © Frank Ho, Amanda Ho, All rights reserved.   www.homathchess.com

Fill in ☐ with answer.

Times	Grouping	Addition
$4 \times \$5 = \square$	4 of $\square$ = 20	$\$5 + \$5 + \$5 + \$5 = \square$
$5 \times \$4 = \square$	5 of $\square$ = 20	$\$4 + \$4 + \$4 + \$4 + \$4 = \square$

Times	Grouping	Addition
$4 \times \$6 = \square$	4 of $\square$ = 24	$\$6 + \$6 + \$6 + \$6 = \square$
$6 \times \$4 = \square$	6 of $\square$ = 24	$\$4 + \$4 + \$4 + \$4 + \$4 + \$4 = \square$

$4 \times 5 = \square = ♖ \times 4 = \square$	$♖ \times 4 = \square = 4 \times 5 = \square$
$4 \times \square = 24 = 6 \times \square = \square$	$6 \times \square = 24 = 6 \times \square = \square$

4 1 ☐	4 5 ☐	4 9 ☐	4 4 ☐	4 8 ☐
4 2 ☐	4 6 ☐	4 1 ☐	4 ♖ ☐	4 9 ☐
4 ♗ ☐	4 7 ☐	4 2 ☐	4 6 ☐	4 ♙ ☐
4 4 ☐	4 8 ☐	4 3 ☐	4 7 ☐	4 2 ☐

Ho Math Chess 何数棋谜   妈!我会棋谜式乘法啦!
Mom! I Learn Multiplication Using Math-Chess-Puzzles Connection!

Student's Name _____ Date _____

2007 - 2018 © Frank Ho, Amanda Ho, All rights reserved.   www.homathchess.com

Fill in _____ and ☐ with answers.

Times	Grouping	Addition
$4 \times \$7 = \square$	4 of $\square = 28$	$\$7 + \$7 + \$7 + \$7 = \square$
$7 \times \$4 = \square$	7 of $\square = 28$	$\$4 + \$4 + \$4 + \$4 + \$4 + \$4 + \$4 = \square$

Times	Grouping	Addition
$4 \times \$8 = \square$	4 of $\square = 32$	$\$8 + \$8 + \$8 + \$8 = \square$
$8 \times \$4 = \square$	8 of $\square = 32$	$\$4 + \$4 + \$4 + \$4 + \$4 + \$4 + \$4 + \$4 = \square$

$4 \times 7 = \square = 7 \times 4 = \square$	$7 \times 4 = \square = 4 \times 7 = \square$
$4 \times \square = 32 = 8 \times \square = \square$	$8 \times \square = 32 = 4 \times \square = \square$

$4\ 1\ \square$	$4\ 5\ \square$	$4\ 9\ \square$	$4\ 4\ \square$	$4\ 8\ \square$
$4\ 2\ \square$	$4\ 6\ \square$	$4\ 1\ \square$	$4\ ♖\ \square$	$4\ 9\ \square$
$4\ 3\ \square$	$4\ 7\ \square$	$4\ 2\ \square$	$4\ 6\ \square$	$4\ 1\ \square$
$4\ 4\ \square$	$4\ 8\ \square$	$4\ ♗\ \square$	$4\ 7\ \square$	$4\ 2\ \square$

Fill in _____ and ☐ with answers.

Times	Grouping	Addition
4 × $9 = ☐	4 of ☐ = 36	$9 + $9 + $9 + $9 = ☐
9 × $4 = ☐	9 of ☐ = 36	$4 + $4 + $4+ $4 + $4 + $4 + $4 + $4 + $4 = ☐

Times	Grouping	Addition
4 × $8 = ☐	4 of ☐ = 32	$8 + $8 + $8 + $8 = ☐
8 × $4 = ☐	8 of ☐ = 32	$4 + $4 + $4 + $4 + $4 + $4 + $4 + $4 = ☐

4 × 9 = ☐ = ♕ × 4 = ☐	9 × 4 = ☐ = 4 × ♕ = ☐
4 × ☐ = 36 = 9 × ☐ = ☐	9 × ☐ = 36 = 4 × ☐ = ☐

4 1 ☐	4 5 ☐	4 ♕ ☐	4 4 ☐	4 8 ☐
4 2 ☐	4 6 ☐	4 1 ☐	4 5 ☐	4 ♕ ☐
4 ♗ ☐	4 7 ☐	4 2 ☐	4 6 ☐	4 1 ☐
4 4 ☐	4 8 ☐	4 ♘ ☐	4 7 ☐	4 2 ☐

2007 - 2018 © Frank Ho, Amanda Ho, All rights reserved.　www.homathchess.com

Preparing for division

☐	☐	☐	☐	☐
X 2	X 3	X 4	X 2	X 4
8	1 2	1 6	8	2 4

☐	☐	☐	☐	☐
X 9	X 8	X 4	X ♞	X 4
3 6	3 2	2 8	1 2	1 6

☐	☐	☐	☐	☐
X ♕	X 4	X 4	X 4	X 7
3 6	2 0	2 8	2 0	2 8

☐	☐	☐	☐	☐
X 5	X ♞	X 4	X 4	X 2
2 0	1 2	3 2	3 6	8

☐	☐	☐	☐	☐
X 4	X ♖	X 6	X 4	X 3
1 6	2 0	2 4	1 2	1 2

2007 - 2018 © Frank Ho, Amanda Ho, All rights reserved. www.homathchess.com

Preparing for division

☐ X 4 = 4	X ☐ 4)‾4‾	☐)4 X 4
☐ X 4 = 8	X ☐ 4)‾8‾	☐)8 X 4
☐ X 4 = 12	X ☐ 3)‾12‾	☐)12 X 4
☐ X 4 = 16	X ☐ 4)‾16‾	☐)16 X 4
☐ X 4 = 20	X ☐ 4)‾20‾	☐)20 X 4
☐ X 4 = 24	X ☐ 4)‾24‾	☐)24 X 4
☐ X 4 = 28	X ☐ 4)‾28‾	☐)28 X 4

Ho Math Chess   何数棋谜   妈!我会棋谜式乘法啦!
Mom! I Learn Multiplication Using Math-Chess-Puzzles Connection!

Student's Name _____ Date _____

2007 - 2018 © Frank Ho, Amanda Ho, All rights reserved.   www.homathchess.com

Preparing for division

☐ X 4 = 32	X ☐ 4) 32	☐) 32 X 4
☐ X 4 = 36	X ☐ 4) 36	☐) 36 X 4
☐ X 4 = 12	X ☐ 3) 12	☐) 12 X 4
☐ X 4 = 16	X ☐ 4) 16	☐) 16 X 4
☐ X 4 = 20	X ☐ 4) 20	☐) 20 X 4
☐ X 4 = 24	X ☐ 4) 24	☐) 24 X 4
☐ X 4 = 28	X ☐ 4) 28	☐) 28 X 4

Ho Math Chess 何数棋谜 妈！我会棋谜式乘法啦！
Mom! I Learn Multiplication Using Math-Chess-Puzzles Connection!

Student's Name _____ Date _____

2007 - 2018 © Frank Ho, Amanda Ho, All rights reserved.   www.homathchess.com

Cross multiplication

12 12 $$\frac{6}{2} = \frac{6}{2}$$	$$\frac{4}{4} = \frac{2}{2}$$	$$\frac{4}{4} = \frac{3}{3}$$	$$\frac{4}{4} = \frac{4}{4}$$
$$\frac{4}{4} = \frac{5}{5}$$	$$\frac{4}{4} = \frac{6}{6}$$	$$\frac{4}{4} = \frac{7}{7}$$	$$\frac{4}{4} = \frac{9}{9}$$
$$\frac{4}{4} = \frac{5}{5}$$	$$\frac{4}{4} = \frac{4}{4}$$	$$\frac{4}{4} = \frac{7}{7}$$	$$\frac{4}{4} = \frac{8}{8}$$
$$\frac{4}{4} = \frac{6}{6}$$	$$\frac{4}{4} = \frac{8}{8}$$	$$\frac{4}{4} = \frac{9}{9}$$	$$\frac{4}{4} = \frac{3}{3}$$

2007 - 2018 © Frank Ho, Amanda Ho, All rights reserved.   www.homathchess.com

Different ways of writing multiplication (Learning division while doing multiplications)

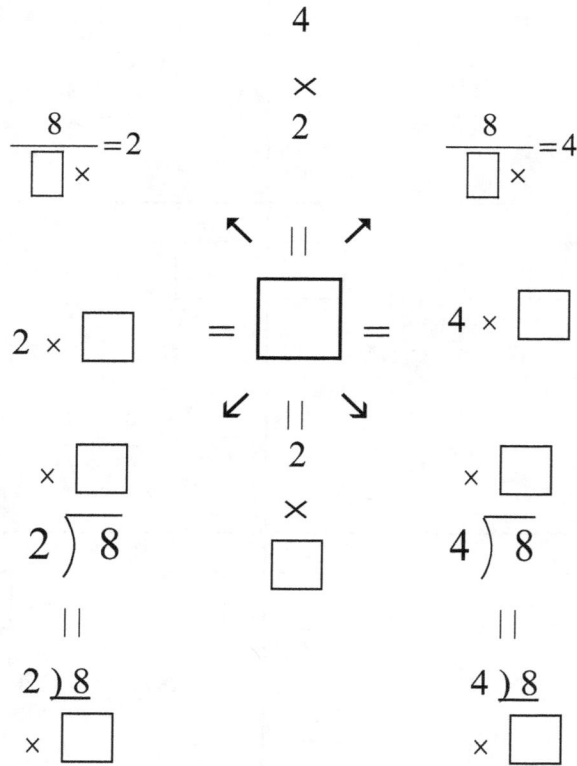

$$4$$
$$\times$$
$$2$$

$$\frac{8}{\boxed{}\times} = 2 \qquad\qquad \frac{8}{\boxed{}\times} = 4$$

$$\nwarrow \; || \; \nearrow$$

$$2 \times \boxed{} \;=\; \boxed{} \;=\; 4 \times \boxed{}$$

$$\swarrow \; || \; \searrow$$

$$\times \boxed{} \qquad 2 \qquad \times \boxed{}$$

$$2\,\overline{)\,8} \qquad \times \qquad 4\,\overline{)\,8}$$
$$\boxed{}$$

$$|| \qquad\qquad\qquad ||$$

$$2\,\underline{)\,8} \qquad\qquad 4\,\underline{)\,8}$$
$$\times \boxed{} \qquad\qquad \times \boxed{}$$

Student's Name _____ Date _____

2007 - 2018 © Frank Ho, Amanda Ho, All rights reserved.   www.homathchess.com

Different ways of writing multiplication (Learning division while doing multiplications)

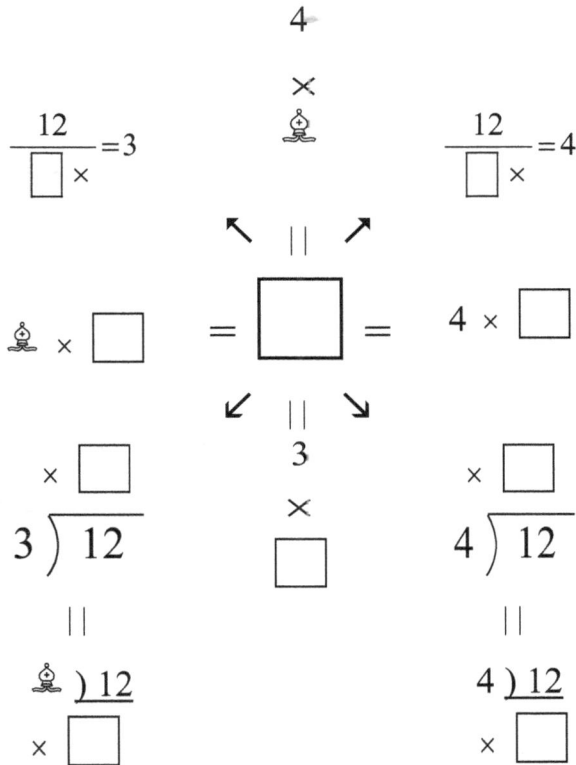

2007 - 2018 © Frank Ho, Amanda Ho, All rights reserved.　www.homathchess.com

Different ways of writing multiplication (Learning division while doing multiplications)

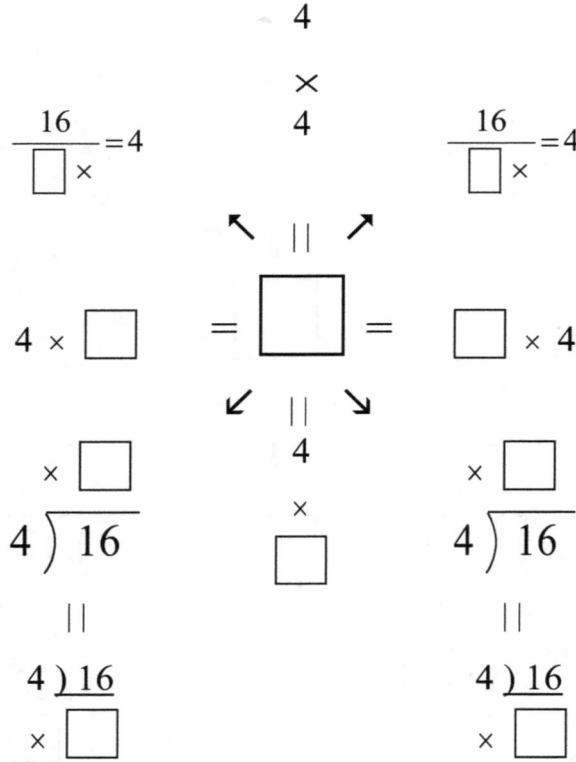

2007 - 2018 © Frank Ho, Amanda Ho, All rights reserved.　　www.homathchess.com

Different ways of writing multiplication (Learning division while doing multiplications)

2007 - 2018 © Frank Ho, Amanda Ho, All rights reserved.　　www.homathchess.com

Different ways of writing multiplication (Learning division while doing multiplications)

Mom! I Learn Multiplication Using Math-Chess-Puzzles Connection!

Student's Name _____ Date _____

2007 - 2018 © Frank Ho, Amanda Ho, All rights reserved.   www.homathchess.com

Different ways of writing multiplication (Learning division while doing multiplications)

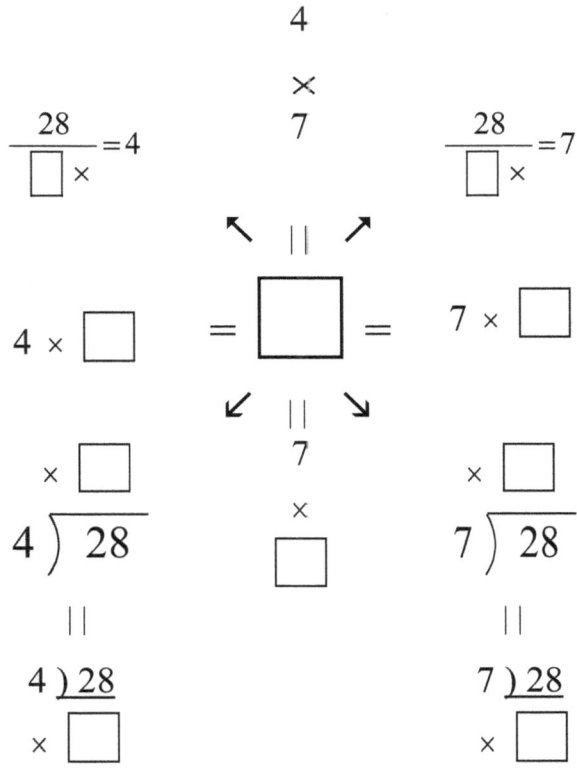

$$\frac{28}{\square \times} = 4 \qquad \qquad \frac{28}{\square \times} = 7$$

$$\begin{array}{c} 4 \\ \times \\ 7 \end{array}$$

$$4 \times \square \qquad = \boxed{} = \qquad 7 \times \square$$

$$\begin{array}{c} 7 \\ \times \\ \square \end{array}$$

$$\begin{array}{c} \times \square \\ 4 \overline{)\,28} \end{array} \qquad \qquad \begin{array}{c} \times \square \\ 7 \overline{)\,28} \end{array}$$

$$4 \underline{)\,28} \qquad \qquad 7 \underline{)\,28}$$
$$\times \square \qquad \qquad \times \square$$

2007 - 2018 © Frank Ho, Amanda Ho, All rights reserved.   www.homathchess.com

Different ways of writing multiplication (Learning division while doing multiplications)

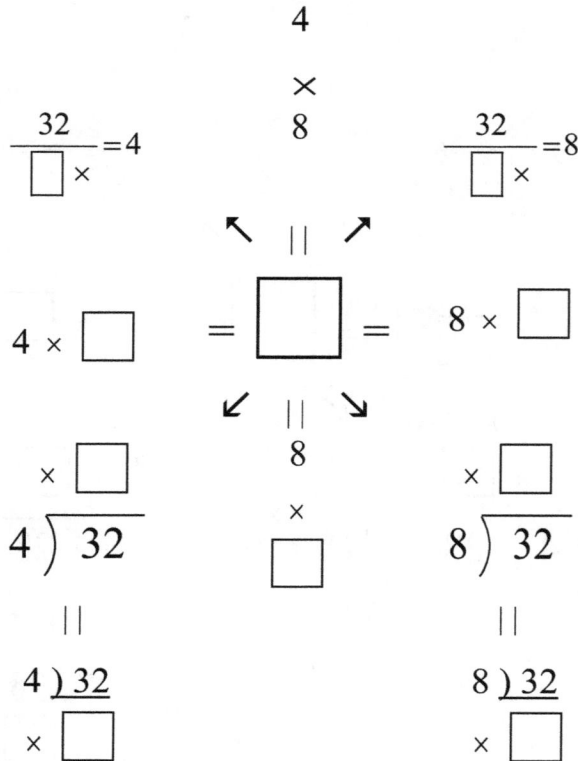

$$4$$
$$\times$$
$$8$$

$$\frac{32}{\boxed{}\times}=4 \qquad\qquad \frac{32}{\boxed{}\times}=8$$

↖ ‖ ↗

$$4 \times \boxed{} \quad = \quad \boxed{} \quad = \quad 8 \times \boxed{}$$

↙ ‖ ↘

$$8$$
$$\times$$
$$\boxed{}$$

$$\begin{array}{r}\times\ \boxed{}\\ 4\,\overline{)\,32}\end{array} \qquad\qquad \begin{array}{r}\times\ \boxed{}\\ 8\,\overline{)\,32}\end{array}$$

‖   ‖

$$\begin{array}{l}4\,)\,32\\ \times\ \boxed{}\end{array} \qquad\qquad \begin{array}{l}8\,)\,32\\ \times\ \boxed{}\end{array}$$

2007 - 2018 © Frank Ho, Amanda Ho, All rights reserved.   www.homathchess.com

Different ways of writing multiplication (Learning division while doing multiplications)

2007 - 2018 © Frank Ho, Amanda Ho, All rights reserved.　　www.homathchess.com

Counting 5's multiples

Write 5's multiples.

♖, 10, 15, ☐,☐,☐,☐,☐,☐

Fill in the following ☐ with a number.

Sequence	♙	2	♗	4	♖	6	7	8	♕
Add 5	☐	10	☐	20	☐	30	☐	40	☐

Sequence	1	2	3	4	5	6	7	8	♕
Add 5	♖	☐	15	☐	25	☐	35	☐	45

Sequence	♙	2	♗	4	5	6	7	8	9
Add 5	☐	10	☐	20	☐	30	☐	40	☐

Sequence	1	2	3	4	5	6	7	8	♕
Add 5	5	☐	15	☐	25	☐	35	☐	45

Sequence	♙	2	3	4	♖	6	7	8	9
Add 5	☐	10	☐	20	☐	30	☐	40	☐

Sequence	1	2	♗	4	5	6	7	8	♕
Add 5	5	☐	15	☐	25	☐	35	☐	45

Mom! I Learn Multiplication Using Math-Chess-Puzzles Connection!

Student's Name _____ Date _____

2007 - 2018 © Frank Ho, Amanda Ho, All rights reserved.   www.homathchess.com

5 times

5 × 2 = ☐	Five times two is ☐	2 × 5 = ☐	two times five is ☐
5 × 3 = ☐	Five times three is ☐	3 × ♖ = ☐	Three times five is ☐
5 × 4 = ☐	Five times four is ☐	4 × 5 = ☐	Four times five is ☐
♖ × 5 = ☐	Five times five is ☐	5 × ♖ = ☐	Five times five is ☐
5 × 6 = ☐	Five times six is ☐	6 × 5 = ☐	Six times five is ☐
♖ × 7 = ☐	Five times seven is ☐	7 × 5 = ☐	Seven times five is ☐
5 × 8 = ☐	Five times eight is ☐	8 × ♖ = ☐	Eight times five is ☐
♖ × 9 = ☐	Five times nine is ☐	9 × 5 = ☐	Nine times five is ☐

♖	1	2	♖	♘
X 1	X 5	X 5	X 2	X 5
☐	☐	☐☐	☐☐	☐☐

5	5	♖	5	5
X 5	X 6	X 7	X 8	X 9
☐☐	☐☐	☐☐	☐☐	☐☐

4	♘	5	7	5
X 5	X ♖	X 5	X 5	X ♛
☐☐	☐☐	☐☐	☐☐	☐☐

Ho Math Chess 何数棋谜 妈！我会棋谜式乘法啦！

Mom! I Learn Multiplication Using Math-Chess-Puzzles Connection!

Student's Name _____ Date _____

2007 - 2018 © Frank Ho, Amanda Ho, All rights reserved.   www.homathchess.com

♖
X 1
□

1
X 5
□

5
X 2
☐☐

2
X 5
☐☐

5
X 3
☐☐

5
X 4
☐☐

5
X 5
☐☐

7
X 5
☐☐

♖
X 8
☐☐

9
X 5
☐☐

5
X 6
☐☐

♖
X 8
☐☐

5
X 7
☐☐

5
X 6
☐☐

♕
X ♖
☐☐

♖
X ♖
☐☐

5
X 6
☐☐

7
X ♖
☐☐

♖
X 8
☐☐

♕
X 5
☐☐

7
X ♖
☐☐

5
X 8
☐☐

♖
X 2
☐☐

5
X 6
☐☐

5
X 9
☐☐

Ho Math Chess 何数棋谜 妈！我会棋谜式乘法啦！
Mom! I Learn Multiplication Using Math-Chess-Puzzles Connection!

Student's Name _____ Date _____

2007 - 2018 © Frank Ho, Amanda Ho, All rights reserved.   www.homathchess.com

Oral practice

five one five	5 1 ☐	5 1 × ♖ ☐ ☐ ☐
five two ten	♖ 2 ☐	5 2 × 5 ☐ ☐ ☐
five three fifteen	5 3 ☐	5 3 × ♖ ☐ ☐ ☐
five four twenty	5 4 ☐	5 4 × 5 ☐ ☐ ☐
five five twenty-five	♖ 5 ☐	2 5 5 × 5 ☐ ☐ ☐

Ho Math Chess   何数棋谜   妈！我会棋谜式乘法啦！
Mom! I Learn Multiplication Using Math-Chess-Puzzles Connection!

Student's Name _____ Date _____
2007 - 2018 © Frank Ho, Amanda Ho, All rights reserved.   www.homathchess.com

Oral practice

five six thirty	♖ 6 ☐	³5 6 × 5 ☐☐☐
five seven thirty-five	5 7 ☐	5 7 × ♖ ☐☐☐
five eight forty	♖ 8 ☐	5 8 × 5 ☐☐☐
five nine forty-five	5 9 ☐	5 9 × 5 ☐☐☐
five five twenty-five	♖ 5 ☐	5 5 × 5 ☐☐☐
five six thirty	5 6 ☐	5 6 × ♖ ☐☐☐

Student's Name _____ Date _____

2007 - 2018 © Frank Ho, Amanda Ho, All rights reserved.　　www.homathchess.com

Fill in ☐ with answer.

Times	Grouping	Addition
$5 \times \$1 = $ ☐	5 of ☐ = 5	$\$1 + \$1 + \$1 + \$1 + \$1 = $ ☐
$1 \times \$5 = $ ☐	1 of ☐ = 5	5 of $\$1 = $ ☐

Fill in ☐ with answer.

Expression	Grouping	Addition
$5 \times \$2$	5 of ☐ = 10	$\$2 + \$2 + \$2 + \$2 + \$2 = $ ☐
$2 \times \$5$	2 of ☐ = 10	$\$5 + \$5 = $ ☐

$5 \times ♙ = $ ☐ $ = ♙ \times 5 = $ ☐	$♙ \times ♜ = $ ☐ $ = 5 \times 1 = $ ☐
$5 \times $ ☐ $ = 10 = 2 \times $ ☐ $ = $ ☐	$2 \times $ ☐ $ = 10 = 5 \times $ ☐ $ = $ ☐

5 1 ☐	5 5 ☐	5 9 ☐	5 4 ☐	5 8 ☐
♜ 2 ☐	5 6 ☐	♜ ♙ ☐	5 5 ☐	5 9 ☐
5 3 ☐	5 7 ☐	5 2 ☐	5 6 ☐	5 1 ☐
5 4 ☐	5 8 ☐	5 3 ☐	5 7 ☐	♜ 2 ☐

Ho Math Chess   何数棋谜   妈!我会棋谜式乘法啦!

Mom! I Learn Multiplication Using Math-Chess-Puzzles Connection!

Student's Name _____ Date _____

2007 - 2018 © Frank Ho, Amanda Ho, All rights reserved.   www.homathchess.com

Fill in ☐ with answer.

Times	Grouping	Addition
5 × \$3 = ☐	5 of ☐ = 15	\$3 + \$3 + \$3 + \$3 + \$3 = ☐
3 × \$5 = ☐	3 of ☐ = 15	\$5 + \$5 + \$5 = ☐

Fill in ☐ with answer.

Expression	Grouping	Addition
5 × \$4	5 of ☐ = 20	\$4 + \$4 + \$4 + \$4 + \$4 = ☐
4 × \$5	4 of ☐ = 20	\$5 + \$5 + \$5 + \$5 = ☐

5 × 3 = ☐ = 3 × 5 = ☐	3 × ♖ = ☐ = 5 × 3 = ☐
♖ × ☐ = 20 = 4 × ☐ = ☐	4 × ☐ = 20 = 5 × ☐ = ☐

5 ♙ ☐	5 5 ☐	5 9 ☐	5 4 ☐	♖ 8 ☐
5 2 ☐	5 6 ☐	♖ 1 ☐	5 5 ☐	♖ 9 ☐
5 3 ☐	♖ 7 ☐	5 2 ☐	♖ 6 ☐	5 ♙ ☐
♖ 4 ☐	5 8 ☐	5 3 ☐	5 7 ☐	5 2 ☐

Ho Math Chess 何数棋谜 妈!我会棋谜式乘法啦!

Mom! I Learn Multiplication Using Math-Chess-Puzzles Connection!

Student's Name _____ Date _____

2007 - 2018 © Frank Ho, Amanda Ho, All rights reserved. www.homathchess.com

Fill in □ with answer.

Times	Grouping	Addition
$5 \times \$5 = \square$	♜ of $\square = 25$	$\$5 + \$5 + \$5 + \$5 + \$5 = \square$
$5 \times \$5 = \square$	5 of $\square = 25$	$\$5 + \$5 + \$5 + \$5 + \$5 = \square$

Times	Grouping	Addition
$5 \times \$6 = \square$	5 of $\square = 30$	$\$6 + \$6 + \$6 + \$6 + \$6 = \square$
$6 \times \$5 = \square$	6 of $\square = 30$	$\$5 + \$5 + \$5 + \$5 + \$5 + \$5 = \square$

$5 \times 5 = \square = 5 \times 5 = \square$	$5 \times 5 = \square = 5 \times 5 = \square$
$5 \times \square = 30 = 6 \times \square = \square$	$6 \times \square = 30 = 5 \times \square = \square$

5 1 □	5 5 □	5 9 □	5 4 □	♜ 8 □
♜ 2 □	5 6 □	5 1 □	♜ 5 □	5 9 □
5 3 □	♜ 7 □	5 2 □	5 6 □	5 1 □
5 4 □	5 8 □	5 3 □	5 7 □	♜ 2 □

2007 - 2018 © Frank Ho, Amanda Ho, All rights reserved.   www.homathchess.com

Fill in _____ and ☐ with answers.

Times	Grouping	Addition
$5 \times \$7 = \square$	$5 \text{ of } \square = 35$	$\$7 + \$7 + \$7 + \$7 + \$7 = \square$
$7 \times \$5 = \square$	$7 \text{ of } \square = 35$	$\$5 + \$5 + \$5 + \$5 + \$5 + \$5 + \$5 = \square$

Times	Grouping	Addition
♖ $\times \$8 = \square$	$5 \text{ of } \square = 40$	$\$8 + \$8 + \$8 + \$8 + \$8 = \square$
$8 \times \$$♖ $= \square$	$8 \text{ of } \square = 40$	$\$5 + \$5 + \$5 + \$5 + \$5 + \$5 + \$5 + \$5 = \square$

$5 \times 7 = \square = 7 \times 5 = \square$	$7 \times 5 = \square = 5 \times 7 = \square$
$5 \times \square = 40 = 8 \times \square = \square$	$8 \times \square = 40 = 5 \times \square = \square$

5 1 ☐	5 5 ☐	♖ 9 ☐	5 4 ☐	♖ 8 ☐
♖ 2 ☐	5 6 ☐	5 1 ☐	5 5 ☐	5 9 ☐
5 3 ☐	♖ 7 ☐	5 2 ☐	♖ 6 ☐	5 1 ☐
5 4 ☐	5 8 ☐	5 3 ☐	5 7 ☐	♖ 2 ☐

Ho Math Chess 何数棋谜 妈！我会棋谜式乘法啦！
Mom! I Learn Multiplication Using Math-Chess-Puzzles Connection!

Student's Name _____ Date _____

2007 - 2018 © Frank Ho, Amanda Ho, All rights reserved.   www.homathchess.com

Fill in _____ and ⬜ with answers.

Times	Grouping	Addition
$5 \times \$9 = \square$	♖ of $\square$ =45	$\$9 + \$9 + \$9 + \$9 + \$9 = \square$
$9 \times \$5 = \square$	9 of $\square$ = 45	$\$5 + \$5 + \$5 + \$5 + \$5 + \$5 + \$5 + \$5 + \$5 = \square$

Times	Grouping	Addition
♖ $\times \$8 = \square$	5 of $\square$ = 40	$\$8 + \$8 + \$8 + \$8 + \$8 = \square$
$8 \times \$5 = \square$	8 of $\square$ = 40	$\$5 + \$5 + \$5 + \$5 + \$5 + \$5 + \$5 + \$5 = \square$

$5 \times 9 = \square = 9 \times ♖ = \square$	$9 \times ♖ = \square = 5 \times 9 = \square$
$5 \times \square = 40 = 8 \times \square = \square$	$8 \times \square = 40 = 5 \times \square = \square$

5 1 ⬜	5 5 ⬜	5 9 ⬜	5 4 ⬜	♖ 8 ⬜
5 2 ⬜	♖ 6 ⬜	5 1 ⬜	♖ 5 ⬜	5 9 ⬜
5 3 ⬜	5 7 ⬜	♖ 2 ⬜	5 6 ⬜	5 1 ⬜
♖ 4 ⬜	5 8 ⬜	5 3 ⬜	5 7 ⬜	♖ 2 ⬜

Student's Name _____ Date _____

2007 - 2018 © Frank Ho, Amanda Ho, All rights reserved.　　www.homathchess.com

Preparing for division

□　　　　□　　　　□　　　　□　　　　□
X　4　　　X　2　　　X　6　　　X　9　　　X　3
20　　　　10　　　　30　　　　45　　　　15

□　　　　□　　　　□　　　　□　　　　□
X　8　　　X　6　　　X　♖　　　X　3　　　X　2
40　　　　30　　　　25　　　　15　　　　10

□　　　　□　　　　□　　　　□　　　　□
X　9　　　X　♖　　　X　7　　　X　5　　　X　7
45　　　　25　　　　35　　　　25　　　　35

□　　　　□　　　　□　　　　□　　　　□
X　8　　　X　6　　　X　♖　　　X　3　　　X　2
40　　　　30　　　　25　　　　15　　　　10

□　　　　□　　　　□　　　　□　　　　□
X　♕　　　X　6　　　X　7　　　X　♖　　　X　7
45　　　　30　　　　35　　　　30　　　　35

2007 - 2018 © Frank Ho, Amanda Ho, All rights reserved.   www.homathchess.com

Preparing for division

□ X ♖ = 5	X □ 5) 5	□) 5 X ♖
□ X 5 = 10	X □ 5) 10	□) 10 X 5
□ X ♖ = 15	X □ 5) 15	□) 15 X ♖
□ X 5 = 20	X □ 5) 20	□) 20 X 5
□ X ♖ = 25	X □ 5) 25	□) 25 X 5
□ X 5 = 30	X □ 5) 30	□) 30 X ♖
□ X 5 = 35	X □ 5) 35	□) 35 X 5

2007 - 2018 © Frank Ho, Amanda Ho, All rights reserved.   www.homathchess.com

Preparing for division

☐ X 5 = 40	X ☐ / 5)40	☐)40 / X 5
☐ X 5 = 45	X ☐ / 5)45	☐)45 / X 5
☐ X ♖ = 5	X ☐ / 5)5	☐)5 / X ♖
☐ X 5 = 10	X ☐ / 5)10	☐)10 / X 5
☐ X ♖ = 15	X ☐ / 5)15	☐)15 / X 5
☐ X 5 = 20	X ☐ / 5)20	☐)20 / X ♖
☐ X ♖ = 25	X ☐ / 5)25	☐)25 / X 5

Student's Name _____ Date _____

2007 - 2018 © Frank Ho, Amanda Ho, All rights reserved.   www.homathchess.com

Cross multiplication

12 12 ↖ ↗ $\dfrac{6}{2} = \dfrac{6}{2}$	□ □ ↖ ↗ $\dfrac{5}{5} = \dfrac{2}{2}$	□ □ ↖ ↗ $\dfrac{5}{5} = \dfrac{3}{3}$	□ □ ↖ ↗ $\dfrac{5}{5} = \dfrac{4}{4}$
□ □ ↖ ↗ $\dfrac{5}{5} = \dfrac{5}{5}$	□ □ ↖ ↗ $\dfrac{5}{5} = \dfrac{6}{6}$	□ □ ↖ ↗ $\dfrac{5}{5} = \dfrac{7}{7}$	□ □ ↖ ↗ $\dfrac{5}{5} = \dfrac{9}{9}$
□ □ ↖ ↗ $\dfrac{5}{5} = \dfrac{5}{5}$	□ □ ↖ ↗ $\dfrac{5}{5} = \dfrac{4}{4}$	□ □ ↖ ↗ $\dfrac{5}{5} = \dfrac{7}{7}$	□ □ ↖ ↗ $\dfrac{5}{5} = \dfrac{8}{8}$
□ □ ↖ ↗ $\dfrac{5}{5} = \dfrac{6}{6}$	□ □ ↖ ↗ $\dfrac{5}{5} = \dfrac{8}{8}$	□ □ ↖ ↗ $\dfrac{5}{5} = \dfrac{9}{9}$	□ □ ↖ ↗ $\dfrac{5}{5} = \dfrac{3}{3}$

Ho Math Chess　何数棋谜　妈!我会棋谜式乘法啦!
Mom! I Learn Multiplication Using Math-Chess-Puzzles Connection!

Student's Name _____ Date _____

2007 - 2018 © Frank Ho, Amanda Ho, All rights reserved.　www.homathchess.com

Different ways of writing multiplication (Learning division while doing multiplications)

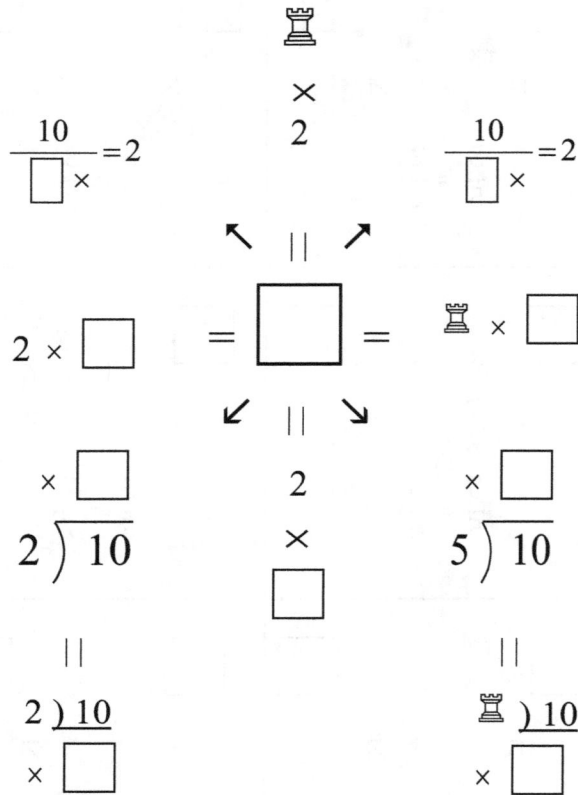

$$\frac{10}{\square \times} = 2 \qquad\qquad \frac{10}{\square \times} = 2$$

♖ × 2

$$2 \times \square \quad = \quad \boxed{} \quad = \quad ♖ \times \square$$

$$2\,)\overline{\,10\,} \times \square \qquad 2 \times \square \qquad 5\,)\overline{\,10\,} \times \square$$

$$2\,)\,10 \qquad\qquad ♖\,)\,10$$
$$\times \square \qquad\qquad \times \square$$

2007 - 2018 © Frank Ho, Amanda Ho, All rights reserved.   www.homathchess.com

Different ways of writing multiplication (Learning division while doing multiplications)

Ho Math Chess 何数棋谜 妈！我会棋谜式乘法啦！
Mom! I Learn Multiplication Using Math-Chess-Puzzles Connection!

Student's Name _____ Date _____
2007 - 2018 © Frank Ho, Amanda Ho, All rights reserved.   www.homathchess.com

Different ways of writing multiplication (Learning division while doing multiplications)

Ho Math Chess　何数棋谜　妈！我会棋谜式乘法啦！
Mom! I Learn Multiplication Using Math-Chess-Puzzles Connection!

Student's Name _____ Date _____

2007 - 2018 © Frank Ho, Amanda Ho, All rights reserved.　　www.homathchess.com

Different ways of writing multiplication (Learning division while doing multiplications)

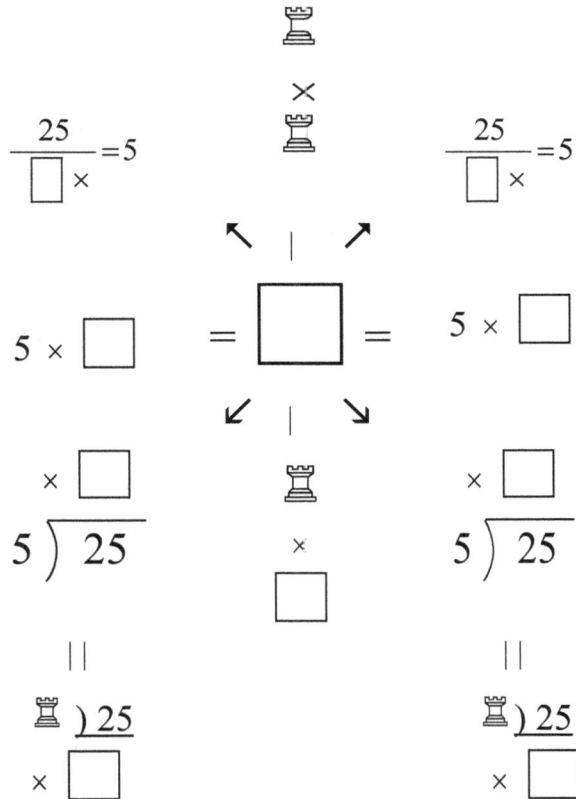

Ho Math Chess   何数棋谜   妈！我会棋谜式乘法啦！
Mom! I Learn Multiplication Using Math-Chess-Puzzles Connection!

Student's Name _____ Date _____

2007 - 2018 © Frank Ho, Amanda Ho, All rights reserved.   www.homathchess.com

Different ways of writing multiplication (Learning division while doing multiplications)

Ho Math Chess 何数棋谜 妈！我会棋谜式乘法啦！
Mom! I Learn Multiplication Using Math-Chess-Puzzles Connection!

Student's Name _____ Date _____

2007 - 2018 © Frank Ho, Amanda Ho, All rights reserved.   www.homathchess.com

Different ways of writing multiplication (Learning division while doing multiplications)

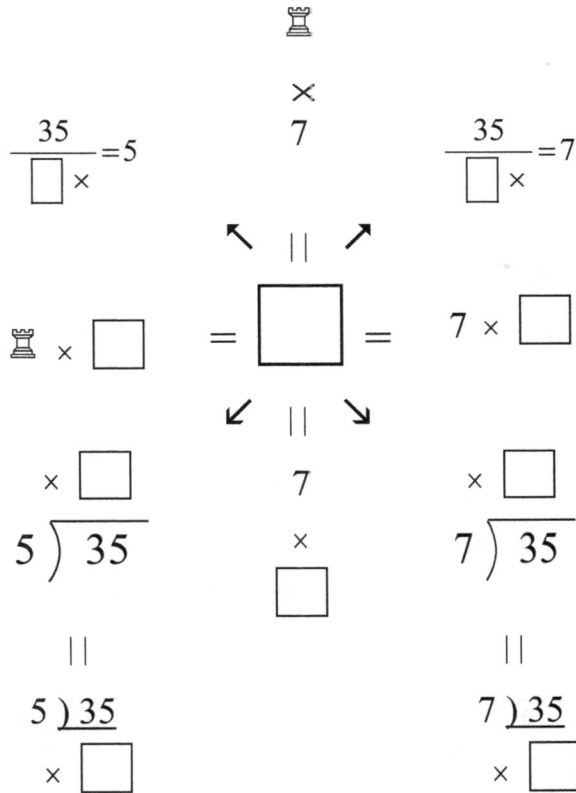

$$\frac{35}{\boxed{}\times} = 5 \qquad \begin{array}{c}\text{♖}\\ \times\\ 7\end{array} \qquad \frac{35}{\boxed{}\times} = 7$$

$$\text{♖} \times \boxed{} \; = \; \boxed{} \; = \; 7 \times \boxed{}$$

$$\times \boxed{} \qquad\qquad 7 \qquad\qquad \times \boxed{}$$

$$5 \overline{)\,35} \qquad\qquad \times \qquad\qquad 7 \overline{)\,35}$$

$$\qquad\qquad\qquad \boxed{}$$

$$\| \qquad\qquad\qquad\qquad \|$$

$$5 \,\underline{)\,35} \qquad\qquad\qquad 7 \,\underline{)\,35}$$

$$\times \boxed{} \qquad\qquad\qquad \times \boxed{}$$

Ho Math Chess   何数棋谜   妈！我会棋谜式乘法啦！
Mom! I Learn Multiplication Using Math-Chess-Puzzles Connection!

Student's Name _____ Date _____

2007 - 2018 © Frank Ho, Amanda Ho, All rights reserved.   www.homathchess.com

Different ways of writing multiplication (Learning division while doing multiplications)

$$\frac{40}{\square\times}=5 \qquad \frac{40}{\square\times}=8$$

$$\square \times \square = \boxed{} = 8 \times \square$$

$$5\,\overline{)\,40} \qquad 8\,\overline{)\,40}$$

Different ways of writing multiplication (Learning division while doing multiplications)

2007 - 2018 © Frank Ho, Amanda Ho, All rights reserved.　www.homathchess.com

Counting 6s multiples

6, 12, 18, ☐,☐,☐,☐,☐,☐

Fill in the following ☐ with a number.

Sequence	♟	2	3	4	5	6	7	8	♛
Add 6	☐	12	☐	24	☐	36	☐	48	☐

Sequence	♟	2	3	4	♜	6	7	8	9
Add 6	6	☐	18	☐	30	☐	42	☐	54

Sequence	1	2	3	4	♜	6	7	8	♛
Add 6	☐	12	☐	24	☐	36	☐	48	☐

Sequence	♟	2	♝	4	5	6	7	8	9
Add 6	6	☐	18	☐	30	☐	42	☐	54

Sequence	1	2	♝	4	5	6	7	8	♛
Add 6	☐	12	☐	24	☐	36	☐	48	☐

Sequence	♟	2	3	4	♜	6	7	8	9
Add 6	6	☐	18	☐	30	☐	42	☐	54

Ho Math Chess   何数棋谜   妈！我会棋谜式乘法啦！
Mom! I Learn Multiplication Using Math-Chess-Puzzles Connection!

Student's Name _____ Date _____

2007 - 2018 © Frank Ho, Amanda Ho, All rights reserved.   www.homathchess.com

6 times

6 × 1 = ☐	Six times one is ☐	♙ × 6= ☐	One times six is ☐
6 × 2 = ☐	Six times two is ☐	2 × 6 = ☐	Two times six is ☐
6 × 3 = ☐	Six times three is ☐	3 × 6 = ☐	Three times six is ☐
6 × 4 = ☐	Six times four is ☐	4 × 6 = ☐	Four times six is ☐
6 × 5 = ☐	Six times five is ☐	5 × 6 = ☐	Five times six is ☐
6 × 6 = ☐	Six times six is ☐	6 × 6 = ☐	Six times six is ☐
6 × 7 = ☐	Six times seven is ☐	7 × 6 = ☐	Seven times six is ☐
6 × 8 = ☐	Six times eight is ☐	8 × 6 = ☐	Eight times six is ☐
6 × 9 = ☐	Six times nine is ☐	9 × 6 = ☐	Nine times six is ☐

6   ♙   2   6   ♗
X 1   X 6   X 6   X 2   X 6

5   6   6   8   5
X 6   X 6   X 7   X 6   X 3

4   ♗   ♖   7   6
X 6   X 6   X 6   X 6   X 9

109

Ho Math Chess 何数棋谜   妈！我会棋谜式乘法啦！
Mom! I Learn Multiplication Using Math-Chess-Puzzles Connection!

Student's Name _____ Date _____

2007 - 2018 © Frank Ho, Amanda Ho, All rights reserved.   www.homathchess.com

6	1	6	2	6
X 1	X 6	X 2	X 6	X 3

6	6	7	6	♛
X 4	X 5	X 6	X 8	X 6

6	6	6	6	9
X 6	X 8	X 7	X 6	X 6

6	6	7	6	♛
X ♜	X 6	X 6	X 8	X 6

7	6	5	6	6
X 6	X 8	X 6	X 6	X 9

Student's Name _____ Date _____

2007 - 2018 © Frank Ho, Amanda Ho, All rights reserved.   www.homathchess.com

Oral practice

six one six	6 1 ☐	$\begin{array}{r} 1\ 1 \\ \times\ \ 6 \\ \hline \square\square \end{array}$
six two twelve	6 2 ☐	$\begin{array}{r} 2\ 2 \\ \times\ \ 6 \\ \hline \square\square\square \end{array}$
six three eighteen	6 ♞ ☐	$\begin{array}{r} 3\ 3 \\ \times\ \ 6 \\ \hline \square\square\square \end{array}$
six four twenty-four	6 4 ☐	$\begin{array}{r} 4\ 4 \\ \times\ \ 6 \\ \hline \square\square\square \end{array}$
six five thirty	6 5 ☐	$\begin{array}{r} ^{3} \\ 5\ 5 \\ \times\ \ 6 \\ \hline \square\square\square \end{array}$

2007 - 2018 © Frank Ho, Amanda Ho, All rights reserved.   www.homathchess.com

Oral practice

six six thirty-six	6 6 ☐	^{3} 6 6 × 6 ☐ ☐ ☐
six seven forty-two	6 7 ☐	7 7 × 6 ☐ ☐ ☐
six eight forty-eight	6 8 ☐	8 8 × 6 ☐ ☐ ☐
six nine fifty-four	6 ♛ ☐	9 9 × 6 ☐ ☐ ☐
six five thirty	6 ♜ ☐	5 5 × 6 ☐ ☐ ☐
six six thirty-six	6 6 ☐	6 6 × 6 ☐ ☐ ☐

Ho Math Chess   何数棋谜   妈！我会棋谜式乘法啦！
Mom! I Learn Multiplication Using Math-Chess-Puzzles Connection!

Student's Name _____ Date _____

2007 - 2018 © Frank Ho, Amanda Ho, All rights reserved.   www.homathchess.com

Fill in ☐ with answer.

Times	Grouping	Addition
$6 \times \$1 = $ ☐	6 of ☐ = 6	$\$1 + \$1 + \$1 + \$1 + \$1 + \$1 = $ ☐
$1 \times \$6 = $ ☐	1 of ☐ = 6	6 of $\$1 = $ ☐

Fill in ☐ with answer.

Expression	Grouping	Addition
$6 \times \$2$	6 of ☐ = 12	$\$2 + \$2 + \$2 + \$2 + \$2 + \$2 = $ ☐
$2 \times \$6$	2 of ☐ = 12	$\$6 + \$6 = $ ☐

$6 \times 1 = $ ☐ $= 1 \times 6 = $ ☐	$1 \times 6 = $ ☐ $= 6 \times 1 = $ ☐
$6 \times $ ☐ $= 12 = 2 \times $ ☐ $= $ ☐	$2 \times $ ☐ $= 12 = 6 \times $ ☐ $= $ ☐

6 ♙ ☐	6 5 ☐	6 9 ☐	6 4 ☐	6 8 ☐
6 2 ☐	6 6 ☐	6 1 ☐	6 ♖ ☐	6 ♛ ☐
6 ♗ ☐	6 7 ☐	6 2 ☐	6 6 ☐	6 1 ☐
6 4 ☐	6 8 ☐	6 ♗ ☐	6 7 ☐	6 2 ☐

Ho Math Chess　何数棋谜　妈！我会棋谜式乘法啦！

Mom! I Learn Multiplication Using Math-Chess-Puzzles Connection!

Student's Name _____ Date _____

2007 - 2018 © Frank Ho, Amanda Ho, All rights reserved.　　www.homathchess.com

Fill in ☐ with answer.

Times	Grouping	Addition
6 × \$♗ = ☐	6 of ☐ = 18	\$3 + \$3 + \$3 + \$3 + \$3 + \$3 = ☐
3 × \$6 = ☐	3 of ☐ = 18	\$6 + \$6 + \$6 = ☐

Fill in ☐ with answer.

Expression	Grouping	Addition
6 × \$4	6 of ☐ = 24	\$4 + \$4 + \$4 + \$4 + \$4 = ☐
4 × \$6	4 of ☐ = 24	\$6 + \$6 + \$6 + \$6 = ☐

6 × 3 = ☐ = ♗ × 6 = ☐	3 × 6 = ☐ = 6 × 3 = ☐
6 × ☐ = 24 = 4 × ☐ = ☐	4 × ☐ = 24 = 6 × ☐ = ☐

6 1 ☐	6 5 ☐	6 9 ☐	6 4 ☐	6 8 ☐
6 2 ☐	6 6 ☐	6 1 ☐	6 5 ☐	6 9 ☐
6 ♗ ☐	6 7 ☐	6 2 ☐	6 6 ☐	6 ♙ ☐
6 4 ☐	6 8 ☐	6 3 ☐	6 7 ☐	6 2 ☐

114

2007 - 2018 © Frank Ho, Amanda Ho, All rights reserved.   www.homathchess.com

Fill in □ with answer.

Times	Grouping	Addition
$6 \times \$5 = \square$	6 of $\square$ = 30	$\$5 + \$5 + \$5 + \$5 + \$5 + \$5 = \square$
$5 \times \$6 = \square$	5 of $\square$ = 30	$\$6 + \$6 + \$6 + \$6 + \$6 = \square$

Times	Grouping	Addition
$6 \times \$6 = \square$	6 of $\square$ = 36	$\$6 + \$6 + \$6 + \$6 + \$6 + \$6 = \square$
$6 \times \$6 = \square$	6 of $\square$ = 36	$\$6 + \$6 + \$6 + \$6 + \$6 + \$6 = \square$

$5 \times 6 = \square = 6 \times 5 = \square$	$6 \times 5 = \square = 5 \times 6 = \square$
$6 \times \square = 36 = 6 \times \square = \square$	$6 \times \square = 36 = 6 \times \square = \square$

6 1 □	6 5 □	6 9 □	6 4 □	6 8 □
6 2 □	6 6 □	6 ♙ □	6 5 □	6 9 □
6 ♗ □	6 7 □	6 2 □	6 6 □	6 1 □
6 4 □	6 8 □	6 ♗ □	6 7 □	6 2 □

2007 - 2018 © Frank Ho, Amanda Ho, All rights reserved.   www.homathchess.com

Fill in ☐ with answers.

Times	Grouping	Addition
$6 \times \$7 = $ ☐	6 of ☐ $= 42$	$\$7 + \$7 + \$7 + \$7 + \$7 + \$7 = $ ☐
$7 \times \$6 = $ ☐	7 of ☐ $= 42$	$\$6 + \$6 + \$6 + \$6 + \$6 + \$6 + \$6 = $ ☐

Times	Grouping	Addition
$6 \times \$8 = $ ☐	6 of ☐ $= 48$	$\$8 + \$8 + \$8 + \$8 + \$8 + \$8 = $ ☐
$8 \times \$6 = $ ☐	8 of ☐ $= 48$	$\$6 + \$6 + \$6 + \$6 + \$6 + \$6 + \$6 + \$6 = $ ☐

$6 \times 7 = $ ☐ $= 7 \times 6 = $ ☐	$7 \times 6 = $ ☐ $= 6 \times 7 = $ ☐
$6 \times $ ☐ $= 48 = 8 \times $ ☐ $= $ ☐	$8 \times $ ☐ $= 48 = 6 \times $ ☐ $= $ ☐

6 1 ☐	6 5 ☐	6 9 ☐	6 4 ☐	6 8 ☐
6 2 ☐	6 6 ☐	6 ♙ ☐	6 5 ☐	6 ♛ ☐
6 ♗ ☐	6 7 ☐	6 2 ☐	6 6 ☐	6 1 ☐
6 4 ☐	6 8 ☐	6 3 ☐	6 7 ☐	6 2 ☐

Mom! I Learn Multiplication Using Math-Chess-Puzzles Connection!

Student's Name _____ Date _____

2007 - 2018 © Frank Ho, Amanda Ho, All rights reserved.   www.homathchess.com

Fill in ☐ with answer.

Times	Grouping	Addition
$6 \times \$9 = $ ☐	6 of ☐ = 54	$\$9 + \$9 + \$9 + \$9 + \$9 + \$9 = $ ☐
$9 \times \$6 = $ ☐	9 of ☐ = 54	$\$6 + \$6 + \$6 + \$6 + \$6 + \$6 + \$6 + \$6 + \$6 = $ ☐

Times	Grouping	Addition
$6 \times \$8 = $ ☐	6 of ☐ = 48	$\$8 + \$8 + \$8 + \$8 + \$8 + \$8 = $ ☐
$8 \times \$6 = $ ☐	8 of ☐ = 48	$\$6 + \$6 + \$6 + \$6 + \$6 + \$6 + \$6 + \$6 = $ ☐

$6 \times 9 = $ ☐ $ = 9 \times 6 = $ ☐	$9 \times 6 = $ ☐ $ = 6 \times 9 = $ ☐
$6 \times $ ☐ $ = 48 = 8 \times $ ☐ $ = $ ☐	$8 \times $ ☐ $ = 48 = 6 \times $ ☐ $ = $ ☐

6 ♙ ☐	6 5 ☐	6 9 ☐	6 4 ☐	6 8 ☐
6 2 ☐	6 6 ☐	6 1 ☐	6 5 ☐	6 ♕ ☐
6 3 ☐	6 7 ☐	6 2 ☐	6 6 ☐	6 1 ☐
6 4 ☐	6 8 ☐	6 ♗ ☐	6 7 ☐	6 2 ☐

Ho Math Chess 何数棋谜 妈！我会棋谜式乘法啦！

Mom! I Learn Multiplication Using Math-Chess-Puzzles Connection!

Student's Name _____ Date _____

2007 - 2018 © Frank Ho, Amanda Ho, All rights reserved.   www.homathchess.com

Preparing for division

☐	☐	☐	☐	☐
X ♗	X 2	X 8	X 9	X 5
18	12	48	54	30

☐	☐	☐	☐	☐
X ♖	X 6	X 4	X 8	X 2
30	18	24	48	12

☐	☐	☐	☐	☐
X 4	X 6	X 5	X 6	X 7
24	42	30	36	42

☐	☐	☐	☐	☐
X 4	X ♖	X 9	X 8	X 7
24	30	54	48	42

☐	☐	☐	☐	☐
X 4	X 5	X 3	X ♕	X 6
24	30	18	54	36

Ho Math Chess 何数棋谜 妈！我会棋谜式乘法啦！
Mom! I Learn Multiplication Using Math-Chess-Puzzles Connection!

Student's Name _____ Date _____

2007 - 2018 © Frank Ho, Amanda Ho, All rights reserved.　www.homathchess.com

Preparing for division

□ X 6 = 6	X □ 6) 6	□) 6 X 6
□ X 6 = 12	X □ 6) 12	□) 12 X 6
□ X 6 = 18	X □ 6) 18	□) 18 X 6
□ X 6 = 24	X □ 6) 24	□) 24 X 6
□ X 6 = 30	X □ 6) 30	□) 30 X 6
□ X 6 = 36	X □ 6) 36	□) 36 X 6
□ X 6 = 42	X □ 6) 42	□) 42 X 6

2007 - 2018 © Frank Ho, Amanda Ho, All rights reserved.   www.homathchess.com

Preparing for division

☐ X 6 = 48	X ☐ 6) 48	☐) 48 X 6
☐ X 6 = 54	X ☐ 6) 54	☐) 54 X 6
☐ X 6 = 30	X ☐ 6) 30	☐) 30 X 6
☐ X 6 = 36	X ☐ 6) 36	☐) 36 X 6
☐ X 6 = 42	X ☐ 6) 42	☐) 42 X 6
☐ X 6 = 48	X ☐ 6) 48	☐) 48 X 6
☐ X 6 = 54	X ☐ 6) 54	☐) 54 X 6

Ho Math Chess 何数棋谜 妈!我会棋谜式乘法啦!
Mom! I Learn Multiplication Using Math-Chess-Puzzles Connection!

Student's Name _____ Date _____

2007 - 2018 © Frank Ho, Amanda Ho, All rights reserved.   www.homathchess.com

Cross multiplication

12 12 ↖ ↗ $\frac{6}{2} = \frac{6}{2}$	□ □ ↖ ↗ $\frac{6}{6} = \frac{2}{2}$	□ □ ↖ ↗ $\frac{6}{6} = \frac{3}{3}$	□ □ ↖ ↗ $\frac{6}{6} = \frac{4}{4}$
□ □ ↖ ↗ $\frac{6}{6} = \frac{5}{5}$	□ □ ↖ ↗ $\frac{6}{6} = \frac{6}{6}$	□ □ ↖ ↗ $\frac{6}{6} = \frac{7}{7}$	□ □ ↖ ↗ $\frac{6}{6} = \frac{9}{9}$
□ □ ↖ ↗ $\frac{6}{6} = \frac{5}{5}$	□ □ ↖ ↗ $\frac{6}{6} = \frac{4}{4}$	□ □ ↖ ↗ $\frac{6}{6} = \frac{7}{7}$	□ □ ↖ ↗ $\frac{6}{6} = \frac{8}{8}$
□ □ ↖ ↗ $\frac{6}{6} = \frac{6}{6}$	□ □ ↖ ↗ $\frac{6}{6} = \frac{8}{8}$	□ □ ↖ ↗ $\frac{6}{6} = \frac{9}{9}$	□ □ ↖ ↗ $\frac{6}{6} = \frac{3}{3}$

Ho Math Chess   何数棋谜   妈！我会棋谜式乘法啦！
Mom! I Learn Multiplication Using Math-Chess-Puzzles Connection!

Student's Name _____ Date _____

2007 - 2018 © Frank Ho, Amanda Ho, All rights reserved.   www.homathchess.com

Different ways of writing multiplication (Learning division while doing multiplications)

$$6$$
$$\times$$
$$2$$

$$\frac{12}{\square \times} = 2 \qquad\qquad \frac{12}{\square \times} = 6$$

$$2 \times \square \quad = \quad \boxed{} \quad = \quad 6 \times \square$$

$$\times \boxed{} \qquad\qquad 2 \qquad\qquad \times \boxed{}$$

$$2\,\overline{)\,12} \qquad\qquad \times \qquad\qquad 6\,\overline{)\,12}$$

$$\boxed{}$$

$$2\,\underline{)\,12} \qquad\qquad\qquad\qquad 6\,\underline{)\,12}$$
$$\times \boxed{} \qquad\qquad\qquad\qquad \times \boxed{}$$

Different ways of writing multiplication (Learning division while doing multiplications)

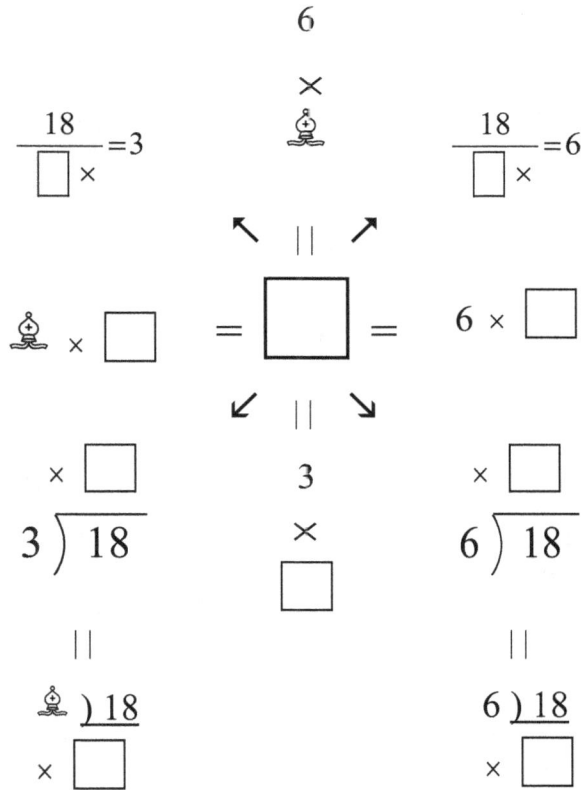

$$6$$
$$\times$$
♗

$$\frac{18}{\boxed{} \times} = 3 \qquad\qquad \frac{18}{\boxed{} \times} = 6$$

↖ ‖ ↗

$$♗ \times \boxed{} \;=\; \boxed{} \;=\; 6 \times \boxed{}$$

↙ ‖ ↘

$$\times \boxed{} \qquad\qquad 3 \qquad\qquad \times \boxed{}$$

$$3\overline{)\,18} \qquad\qquad \times \qquad\qquad 6\overline{)\,18}$$
$$\boxed{}$$

‖ ‖

$$♗\,)\,\underline{18} \qquad\qquad\qquad 6\,)\,\underline{18}$$
$$\times \boxed{} \qquad\qquad\qquad \times \boxed{}$$

123

2007 - 2018 © Frank Ho, Amanda Ho, All rights reserved.   www.homathchess.com

Different ways of writing multiplication (Learning division while doing multiplications)

$$6$$
$$\times$$
$$4$$

$$\frac{24}{\square \times} = 4 \qquad \qquad \frac{24}{\square \times} = 6$$

$$4 \times \square \quad = \quad \boxed{} \quad = \quad \square \times 6$$

$$\times \square \qquad\qquad 4 \qquad\qquad \times \square$$
$$4 \overline{)\, 24} \qquad\qquad \times \qquad\qquad 6 \overline{)\, 24}$$
$$\square$$

$$||$$

$$4 \underline{)\, 24} \qquad\qquad\qquad 6 \underline{)\, 24}$$
$$\times \square \qquad\qquad\qquad \times \square$$

Ho Math Chess 何数棋谜 妈！我会棋谜式乘法啦！
Mom! I Learn Multiplication Using Math-Chess-Puzzles Connection!

Student's Name _____ Date _____
2007 - 2018 © Frank Ho, Amanda Ho, All rights reserved.   www.homathchess.com

Different ways of writing multiplication (Learning division while doing multiplications)

$$\frac{30}{\boxed{}\times} = 5 \qquad 6 \times \text{♜} \qquad \frac{30}{\boxed{}\times} = 6$$

$$\text{♜} \times \boxed{} \ = \ \boxed{} \ = \ 6 \times \boxed{}$$

$$5 \overline{)\,30} \qquad \text{♜} \times \boxed{} \qquad 6 \overline{)\,30}$$

$$\text{♜})\,30 \qquad\qquad 6)\,30$$
$$\times \boxed{} \qquad\qquad \times \boxed{}$$

Student's Name _____ Date _____

2007 - 2018 © Frank Ho, Amanda Ho, All rights reserved.   www.homathchess.com

Different ways of writing multiplication (Learning division while doing multiplications)

$$6 \times 6$$

$$\frac{36}{\Box \times} = 6 \qquad \frac{36}{\Box \times} = 6$$

$$6 \times \Box = \boxed{\ } = 6 \times \Box$$

$$\times \Box \qquad \times \Box$$

$$6)\overline{36} \qquad 6 \times \Box \qquad 6)\overline{36}$$

$$6)\,36 \qquad\qquad 6)\,36$$
$$\times \Box \qquad\qquad \times \Box$$

2007 - 2018 © Frank Ho, Amanda Ho, All rights reserved.   www.homathchess.com

Different ways of writing multiplication (Learning division while doing multiplications)

$$6 \times 7$$

$$\frac{42}{\Box \times} = 6 \qquad\qquad \frac{42}{\Box \times} = 7$$

$$6 \times \Box \qquad = \boxed{} = \qquad 7 \times \Box$$

$$\begin{array}{c}\times\, \Box \\ 6\,\overline{)\,42}\end{array} \qquad \begin{array}{c} 7 \\ \times \\ \Box \end{array} \qquad \begin{array}{c}\times\, \Box \\ 7\,\overline{)\,42}\end{array}$$

$$\begin{array}{c} 6\,\underline{)\,42} \\ \times\, \Box \end{array} \qquad\qquad \begin{array}{c} 7\,\underline{)\,42} \\ \times\, \Box \end{array}$$

2007 - 2018 © Frank Ho, Amanda Ho, All rights reserved.   www.homathchess.com

Different ways of writing multiplication (Learning division while doing multiplications)

$$6 \times 8$$

$$\frac{48}{\Box \times} = 6 \qquad \frac{48}{\Box \times} = 8$$

$$6 \times \Box \quad = \boxed{} = \quad 8 \times \Box$$

$$\times \Box \qquad\qquad 8 \times \Box$$

$$6\overline{)48} \qquad \times \Box \qquad 8\overline{)48}$$

$$|| \qquad\qquad ||$$

$$6\,)\,48 \qquad\qquad 8\,)\,48$$
$$\times \Box \qquad\qquad \times \Box$$

Student's Name _____ Date _____

2007 - 2018 © Frank Ho, Amanda Ho, All rights reserved.　www.homathchess.com

Different ways of writing multiplication (Learning division while doing multiplications)

Mom! I Learn Multiplication Using Math-Chess-Puzzles Connection!

Student's Name _____ Date _____

2007 - 2018 © Frank Ho, Amanda Ho, All rights reserved.　www.homathchess.com

Counting 7's multiples

7, 14, 21, □, □, □, □, □, □

Fill in the following □ with a number.

Sequence	1	2	♘	4	♖	6	7	8	♛
Add 7	□	14	□	28	□	42	□	56	□

Sequence	1	2	3	4	5	6	7	8	9
Add 7	7	□	21	□	35	□	49	□	63

Sequence	1	2	♗	4	♖	6	7	8	♛
Add 7	□	14	□	28	□	42	□	56	□

Sequence	1	2	3	4	5	6	7	8	9
Add 7	7	□	21	□	35	□	49	□	63

Sequence	1	2	♗	4	♖	6	7	8	♛
Add 7	□	14	□	28	□	42	□	56	□

Sequence	1	2	3	4	5	6	7	8	9
Add 7	7	□	21	□	35	□	49	□	63

2007 - 2018 © Frank Ho, Amanda Ho, All rights reserved.   www.homathchess.com

7 times

7 × 1 = ☐	Seven times one is ☐	1 × 7 = ☐	One times seven is ☐
7 × 2 = ☐	Seven times two is ☐	2 × 7 = ☐	Two times seven is ☐
7 × 3 = ☐	Seven times three is ☐	3 × 7= ☐	Three times seven is ☐
7 × 4 = ☐	Seven times four is ☐	4 × 7 = ☐	Four times seven is ☐
7 × 5 = ☐	Seven times five is ☐	5 × 7 = ☐	Five times seven is ☐
7 × 6 = ☐	Seven times six is ☐	6 × 7 = ☐	Six times seven is ☐
7 × 7 = ☐	Seven times seven is ☐	7 × 7 = ☐	Seven times seven is ☐
7 × 8 = ☐	Seven times eight is ☐	8 × 7 = ☐	Eight times seven is ☐
7 × 9 = ☐	Seven times nine is ☐	9 × 7 = ☐	Nine times seven is ☐

$$
\begin{array}{ccccc}
7 & 1 & 7 & 7 & ♗ \\
\times 1 & \times 7 & \times 6 & \times 2 & \times 7 \\
\end{array}
$$

$$
\begin{array}{ccccc}
♖ & 7 & 7 & 8 & 7 \\
\times 7 & \times 6 & \times 7 & \times 7 & \times 3 \\
\end{array}
$$

$$
\begin{array}{ccccc}
4 & 7 & 5 & 7 & 7 \\
\times 7 & \times 6 & \times 7 & \times 7 & \times ♕ \\
\end{array}
$$

	7	1	7	2	7
	X 1	X 7	X 2	X 7	X 3

	6	7	7	7	9
	X 7	X 5	X 7	X 8	X 7

	6	8	7	4	♛
	X 7	X 7	X 7	X 7	X 7

	♜	6	7	8	9
	X 7	X 7	X 3	X 7	X 7

	7	7	♖	7	7
	X 7	X 8	X 7	X 6	X 9

Student's Name _____ Date _____

2007 - 2018 © Frank Ho, Amanda Ho, All rights reserved.　www.homathchess.com

Oral practice

seven one seven	7 1 ☐	1 1 × 7 ☐ ☐
seven two fourteen	7 2 ☐	2 2 × 7 ☐ ☐ ☐
seven three twenty-one	7 ♘ ☐	3 3 × 7 ☐ ☐ ☐
seven four twenty-eight	7 4 ☐	4 4 × 7 ☐ ☐ ☐
seven five thirty-five	7 ♖ ☐	3 5 5 × 7 ☐ ☐ ☐

Mom! I Learn Multiplication Using Math-Chess-Puzzles Connection!

Student's Name _____ Date _____

2007 - 2018 © Frank Ho, Amanda Ho, All rights reserved.   www.homathchess.com

Oral practice

seven six forty-two	7 6 ☐	$\begin{array}{r} 4 \\ 6\,6 \\ \times\ \ 7 \\ \hline \square\square\square \end{array}$
seven seven forty-nine	7 7 ☐	$\begin{array}{r} 7\,7 \\ \times\ \ 7 \\ \hline \square\square\square \end{array}$
seven eight fifty-six	7 8 ☐	$\begin{array}{r} 8\,8 \\ \times\ \ 7 \\ \hline \square\square\square \end{array}$
seven nine sixty-three	7 ♛ ☐	$\begin{array}{r} 9\,9 \\ \times\ \ 7 \\ \hline \square\square\square \end{array}$
seven five thirty-five	7 ♜ ☐	$\begin{array}{r} 5\,5 \\ \times\ \ 7 \\ \hline \square\square\square \end{array}$
two six forty-two	7 6 ☐	$\begin{array}{r} 6\,6 \\ \times\ \ 7 \\ \hline \square\square\square \end{array}$

2007 - 2018 © Frank Ho, Amanda Ho, All rights reserved. www.homathchess.com

Preparing for division

☐	☐	☐	☐	☐
X 2	X 5	X 4	X 6	X 3
14	35	28	42	21

☐	☐	☐	☐	☐
X 8	X 6	X 2	X ♛	X 7
56	42	14	63	49

☐	☐	☐	☐	☐
X 4	X 2	X ♜	X 2	X 7
28	14	35	14	49

☐	☐	☐	☐	☐
X 6	X 5	X 2	X 3	X 4
42	35	14	21	28

☐	☐	☐	☐	☐
X ♜	X 6	X ♝	X 4	X 7
35	42	21	28	49

Student's Name _____ Date _____

2007 - 2018 © Frank Ho, Amanda Ho, All rights reserved.   www.homathchess.com

Fill in ☐ with answer.

Times	Grouping	Addition
$7 \times \$1 = $ ☐	7 of ☐ = 7	$\$1 + \$1 + \$1 + \$1 + \$1 + \$1 + \$1 = $ ☐
$1 \times \$7 = $ ☐	1 of ☐ = 7	7 of $\$1 = $ ☐

Fill in ☐ with answer.

Expression	Grouping	Addition
$7 \times \$2$	7 of ☐ = 14	$\$2 + \$2 + \$2 + \$2 + \$2 + \$2 + \$2 = $ ☐
$2 \times \$7$	2 of ☐ = 14	$\$7 + \$7 = $ ☐

$7 \times 1 = $ ☐ $= 1 \times 7 = $ ☐	$1 \times 7 = $ ☐ $= 7 \times 1 = $ ☐
$7 \times $ ☐ $= 14 = 2 \times $ ☐ $= $ ☐	$2 \times $ ☐ $= 14 = 7 \times $ ☐ $= $ ☐

7 1 ☐	7 5 ☐	7 9 ☐	7 4 ☐	7 8 ☐
7 2 ☐	7 6 ☐	7 1 ☐	7 ♜ ☐	7 ♛ ☐
7 ♝ ☐	7 7 ☐	7 2 ☐	7 6 ☐	7 1 ☐
7 4 ☐	7 8 ☐	7 3 ☐	7 7 ☐	7 2 ☐

Student's Name _____ Date _____

2007 - 2018 © Frank Ho, Amanda Ho, All rights reserved.   www.homathchess.com

Fill in ☐ with answer.

Times	Grouping	Addition
$7 \times \$3 = $ ☐	7 of ☐ $= 21$	$\$3 + \$3 + \$3 + \$3 + \$3 + \$3 + \$3 = $ ☐
$3 \times \$7 = $ ☐	3 of ☐ $= 21$	$\$7 + \$7 + \$7 = $ ☐

Fill in ☐ with answer.

Expression	Grouping	Addition
$7 \times \$4$	7 of ☐ $= 28$	$\$4 + \$4 + \$4 + \$4 + \$4 + \$4 + \$4 = $ ☐
$4 \times \$7$	4 of ☐ $= 28$	$\$7 + \$7 + \$7 + \$7 = $ ☐

$7 \times ♘ = $ ☐ $= 3 \times 7 = $ ☐	$3 \times 7 = $ ☐ $= 7 \times 3 = $ ☐
$7 \times $ ☐ $= 28 = 4 \times $ ☐ $= $ ☐	$4 \times $ ☐ $= 28 = 7 \times $ ☐ $= $ ☐

7 1 ☐	7 5 ☐	7 ♕ ☐	7 4 ☐	7 8 ☐
7 2 ☐	7 6 ☐	7 1 ☐	7 ♖ ☐	7 ♕ ☐
7 ♗ ☐	7 7 ☐	7 2 ☐	7 6 ☐	7 1 ☐
7 4 ☐	7 8 ☐	7 3 ☐	7 7 ☐	7 2 ☐

137

2007 - 2018 © Frank Ho, Amanda Ho, All rights reserved.　　www.homathchess.com

Fill in ☐ with answer.

Times	Grouping	Addition
$7 \times \$5 = $ ☐	7 of ☐ = 35	$\$5 + \$5 + \$5 + \$5 + \$5 + \$5 + \$5 = $ ☐
$5 \times \$7 = $ ☐	5 of ☐ = 35	$\$7 + \$7 + \$7 + \$7 + \$7 = $ ☐

Times	Grouping	Addition
$7 \times \$6 = $ ☐	6 of ☐ = 42	$\$6 + \$6 + \$6 + \$6 + \$6 + \$6 + \$6 = $ ☐
$6 \times \$7 = $ ☐	6 of ☐ = 42	$\$7 + \$7 + \$7 + \$7 + \$7 + \$7 = $ ☐

$7 \times$ ♖ $=$ ☐ $=$ ♖ $\times 7 = $ ☐	$5 \times 7 = $ ☐ $= 7 \times 5 = $ ☐
$7 \times$ ☐ $= 42 = 6 \times$ ☐ $= $ ☐	$6 \times$ ☐ $= 42 = 7 \times$ ☐ $= $ ☐

7 ♙ ☐	7 5 ☐	7 9 ☐	7 4 ☐	7 8 ☐
7 2 ☐	7 6 ☐	7 1 ☐	7 ♖ ☐	7 ♕ ☐
7 3 ☐	7 7 ☐	7 2 ☐	7 6 ☐	7 ♙ ☐
7 4 ☐	7 8 ☐	7 3 ☐	7 7 ☐	7 2 ☐

Ho Math Chess   何数棋谜 妈!我会棋谜式乘法啦!
Mom! I Learn Multiplication Using Math-Chess-Puzzles Connection!

Student's Name _____ Date _____

2007 - 2018 © Frank Ho, Amanda Ho, All rights reserved.   www.homathchess.com

Fill in _____ and ☐ with answers.

Times	Grouping	Addition
$7 \times \$7 = \square$	7 of $\square$ = 49	$\$7 + \$7 + \$7 + \$7 + \$7 + \$7 + \$7 = \square$
$7 \times \$7 = \square$	7 of $\square$ = 49	$\$7 + \$7 + \$7 + \$7 + \$7 + \$7 + \$7 = \square$

Times	Grouping	Addition
$7 \times \$8 = \square$	7 of $\square$ = 56	$\$8 + \$8 + \$8 + \$8 + \$8 + \$8 + \$8 = \square$
$8 \times \$7 = \square$	8 of $\square$ = 56	$\$7 + \$7 + \$7 + \$7 + \$7 + \$7 + \$7 + \$7 = \square$

$7 \times 7 = \square = 7 \times 7 = \square$	$7 \times 7 = \square = 7 \times 7 = \square$
$7 \times \square = 56 = 8 \times \square = \square$	$8 \times \square = 56 = 7 \times \square = \square$

7 ♙ ☐	7 5 ☐	7 9 ☐	7 4 ☐	7 8 ☐
7 2 ☐	7 6 ☐	7 ♙ ☐	7 ♖ ☐	7 ♛ ☐
7 3 ☐	7 7 ☐	7 2 ☐	7 6 ☐	7 1 ☐
7 4 ☐	7 8 ☐	7 ♘ ☐	7 7 ☐	7 2 ☐

Ho Math Chess   何数棋谜   妈!我会棋谜式乘法啦!

Mom! I Learn Multiplication Using Math-Chess-Puzzles Connection!

Student's Name _____ Date _____

2007 - 2018 © Frank Ho, Amanda Ho, All rights reserved.   www.homathchess.com

Fill in [ ] with answer.

Times	Grouping	Addition
$7 \times \$9 = \square$	7 of $\square$ = 63	$\$9 + \$9 + \$9 + \$9 + \$9 + \$9 + \$9 = \square$
$9 \times \$7 = \square$	9 of $\square$ = 63	$\$7 + \$7 + \$7 + \$7 + \$7 + \$7 + \$7 + \$7 + \$7 = \square$

Times	Grouping	Addition
$7 \times \$8 = \square$	7 of $\square$ = 56	$\$8 + \$8 + \$8 + \$8 + \$8 + \$8 + \$8 = \square$
$8 \times \$7 = \square$	8 of $\square$ = 56	$\$7 + \$7 + \$7 + \$7 + \$7 + \$7 + \$7 + \$7 = \square$

$7 \times 9 = \square = ♛ \times 7 = \square$	$♛ \times 7 = \square = 7 \times 9 = \square$
$9 \times \square = 63 = 9 \times \square = \square$	$7 \times \square = 63 = 9 \times \square = \square$

7 ♙ $\square$	7 5 $\square$	7 9 $\square$	7 4 $\square$	7 8 $\square$
7 2 $\square$	7 6 $\square$	7 1 $\square$	7 ♖ $\square$	7 ♛ $\square$
7 ♝ $\square$	7 7 $\square$	7 2 $\square$	7 6 $\square$	7 ♙ $\square$
7 4 $\square$	7 8 $\square$	7 ♝ $\square$	7 7 $\square$	7 2 $\square$

Student's Name _____ Date _____

2007 - 2018 © Frank Ho, Amanda Ho, All rights reserved.　www.homathchess.com

Preparing for division

☐ X 7 = 7	X ☐ 7⟌7	☐ ⟌7 X 7
☐ X 7 = 14	X ☐ 7⟌14	☐ ⟌14 X 7
☐ X 7 = 21	X ☐ 7⟌21	☐ ⟌21 X 7
☐ X 7 = 28	X ☐ 7⟌28	☐ ⟌28 X 7
☐ X 7 = 35	X ☐ 7⟌35	☐ ⟌35 X 7
☐ X 7 = 42	X ☐ 7⟌42	☐ ⟌42 X 7
☐ X 7 = 49	X ☐ 7⟌49	☐ ⟌49 X 7

Ho Math Chess   何数棋谜   妈!我会棋谜式乘法啦!
Mom! I Learn Multiplication Using Math-Chess-Puzzles Connection!

Student's Name _____ Date _____

2007 - 2018 © Frank Ho, Amanda Ho, All rights reserved.   www.homathchess.com

Preparing for division

☐ X 7 = 56	X ☐ 7)56	☐)56 X 7
☐ X 7 = 63	X ☐ 7)63	☐)63 X 7
☐ X 7 = 7	X ☐ 7)7	☐)7 X 7
☐ X 7 = 14	X ☐ 7)14	☐)14 X 7
☐ X 7 = 21	X ☐ 7)21	☐)21 X 7
☐ X 7 = 28	X ☐ 7)28	☐)28 X 7
☐ X 7 = 35	X ☐ 7)35	☐)35 X 7

Ho Math Chess 何数棋谜 妈!我会棋谜式乘法啦!
Mom! I Learn Multiplication Using Math-Chess-Puzzles Connection!

Student's Name _____ Date _____

2007 - 2018 © Frank Ho, Amanda Ho, All rights reserved.   www.homathchess.com

Cross multiplication

12 12 ↖ ↗ $\dfrac{6}{2} = \dfrac{6}{2}$	□ □ ↖ ↗ $\dfrac{7}{7} = \dfrac{2}{2}$	□ □ ↖ ↗ $\dfrac{7}{7} = \dfrac{3}{3}$	□ □ ↖ ↗ $\dfrac{7}{7} = \dfrac{4}{4}$
□ □ ↖ ↗ $\dfrac{7}{7} = \dfrac{5}{5}$	□ □ ↖ ↗ $\dfrac{7}{7} = \dfrac{6}{6}$	□ □ ↖ ↗ $\dfrac{7}{7} = \dfrac{7}{7}$	□ □ ↖ ↗ $\dfrac{7}{7} = \dfrac{9}{9}$
□ □ ↖ ↗ $\dfrac{7}{7} = \dfrac{5}{5}$	□ □ ↖ ↗ $\dfrac{7}{7} = \dfrac{4}{4}$	□ □ ↖ ↗ $\dfrac{7}{7} = \dfrac{7}{7}$	□ □ ↖ ↗ $\dfrac{7}{7} = \dfrac{8}{8}$
□ □ ↖ ↗ $\dfrac{7}{7} = \dfrac{6}{6}$	□ □ ↖ ↗ $\dfrac{7}{7} = \dfrac{8}{8}$	□ □ ↖ ↗ $\dfrac{7}{7} = \dfrac{9}{9}$	□ □ ↖ ↗ $\dfrac{7}{7} = \dfrac{3}{3}$

2007 - 2018 © Frank Ho, Amanda Ho, All rights reserved.   www.homathchess.com

Different ways of writing multiplication (Learning division while doing multiplications)

$$7 \times 2$$

$$\frac{14}{\Box \times} = 2 \qquad\qquad \frac{14}{\Box \times} = 2$$

$$2 \times \Box \quad = \quad \boxed{} \quad = \quad 7 \times \Box$$

$$\times \Box \qquad\qquad 2 \qquad\qquad \times \Box$$

$$2\overline{)\,14} \qquad\qquad \times \qquad\qquad 7\overline{)\,14}$$

$$\Box$$

$$2\,)\,14 \qquad\qquad\qquad 7\,)\,14$$

$$\times \Box \qquad\qquad\qquad\qquad \times \Box$$

2007 - 2018 © Frank Ho, Amanda Ho, All rights reserved.   www.homathchess.com

Different ways of writing multiplication (Learning division while doing multiplications)

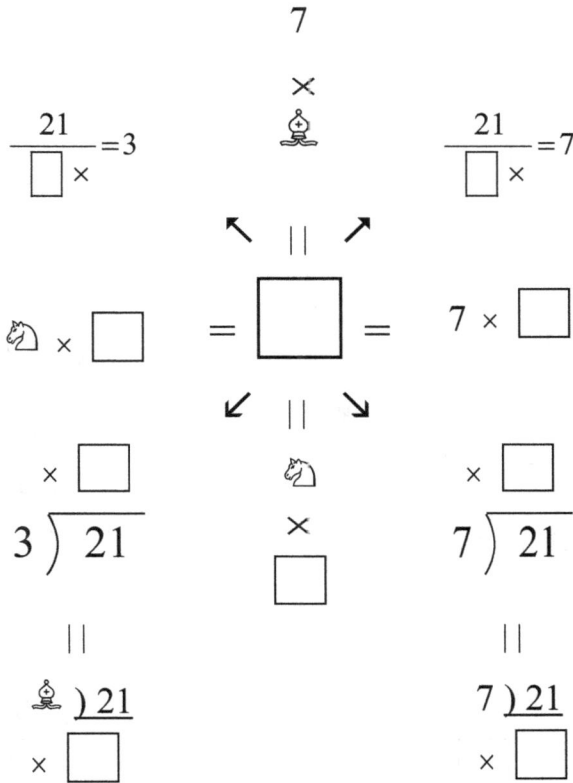

$$\frac{21}{\square \times} = 3 \qquad\qquad \frac{21}{\square \times} = 7$$

$$\square \times \square = \boxed{} = 7 \times \square$$

$$3\overline{)\,21} \qquad ♘ \times \qquad 7\overline{)\,21}$$

$$♗\,\overline{)\,21} \qquad\qquad 7\,\overline{)\,21}$$

2007 - 2018 © Frank Ho, Amanda Ho, All rights reserved.   www.homathchess.com

Different ways of writing multiplication (Learning division while doing multiplications)

$$7$$
$$\times$$
$$4$$

$$\frac{28}{\boxed{}\times} = 4 \qquad\qquad \frac{28}{\boxed{}\times} = 7$$

$$\nwarrow \quad || \quad \nearrow$$

$$4 \times \boxed{} \quad = \quad \boxed{} \quad = \quad \boxed{} \times 7$$

$$\swarrow \quad || \quad \searrow$$

$$\times \boxed{} \qquad\qquad 4 \qquad\qquad \times \boxed{}$$

$$4\overline{)28} \qquad\qquad \times \qquad\qquad 7\overline{)28}$$

$$\boxed{}$$

$$|| \qquad\qquad\qquad\qquad ||$$

$$4\underline{)28} \qquad\qquad\qquad 7\underline{)28}$$
$$\times \boxed{} \qquad\qquad\qquad \times \boxed{}$$

Ho Math Chess　　何数棋谜　妈！我会棋谜式乘法啦！
Mom! I Learn Multiplication Using Math-Chess-Puzzles Connection!

Student's Name _____ Date _____

2007 - 2018 © Frank Ho, Amanda Ho, All rights reserved.　　www.homathchess.com

Different ways of writing multiplication (Learning division while doing multiplications)

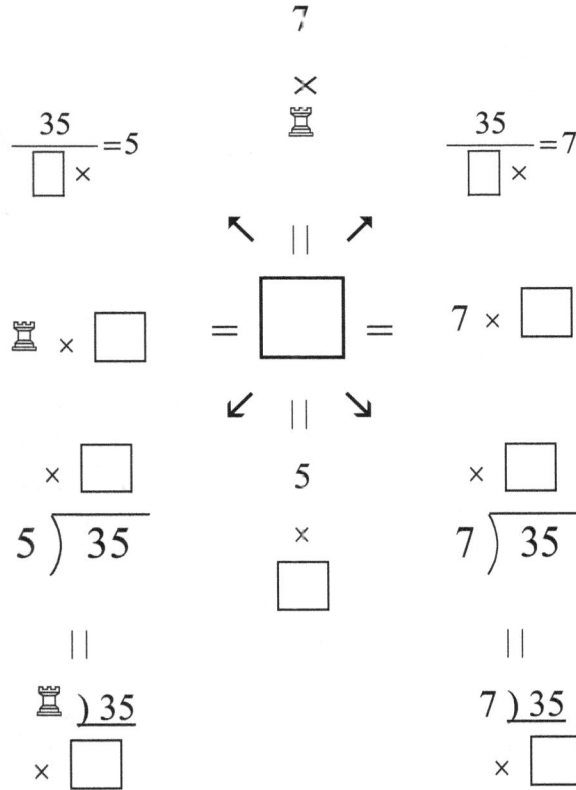

Ho Math Chess   何数棋谜   妈！我会棋谜式乘法啦！
Mom! I Learn Multiplication Using Math-Chess-Puzzles Connection!

Student's Name _____ Date _____

2007 - 2018 © Frank Ho, Amanda Ho, All rights reserved.   www.homathchess.com

Different ways of writing multiplication (Learning division while doing multiplications)

$$7$$
$$\times$$
$$6$$

$$\frac{42}{\boxed{}\times}=6$$
$$\frac{42}{\boxed{}\times}=7$$

↖ || ↗

$$6 \times \boxed{} \quad = \quad \boxed{} \quad = \quad 7 \times \boxed{}$$

↙ || ↘

$$\times \boxed{}$$
$$6$$
$$\times \boxed{}$$

$$6\overline{)\,42}$$ $$\times$$ $$7\overline{)\,42}$$

$$\boxed{}$$

|| ||

$$6\underline{)\,42}$$ $$7\underline{)\,42}$$
$$\times \boxed{}$$ $$\times \boxed{}$$

Ho Math Chess   何数棋谜   妈！我会棋谜式乘法啦！
Mom! I Learn Multiplication Using Math-Chess-Puzzles Connection!

Student's Name _____ Date _____

2007 - 2018 © Frank Ho, Amanda Ho, All rights reserved.   www.homathchess.com

Different ways of writing multiplication (Learning division while doing multiplications)

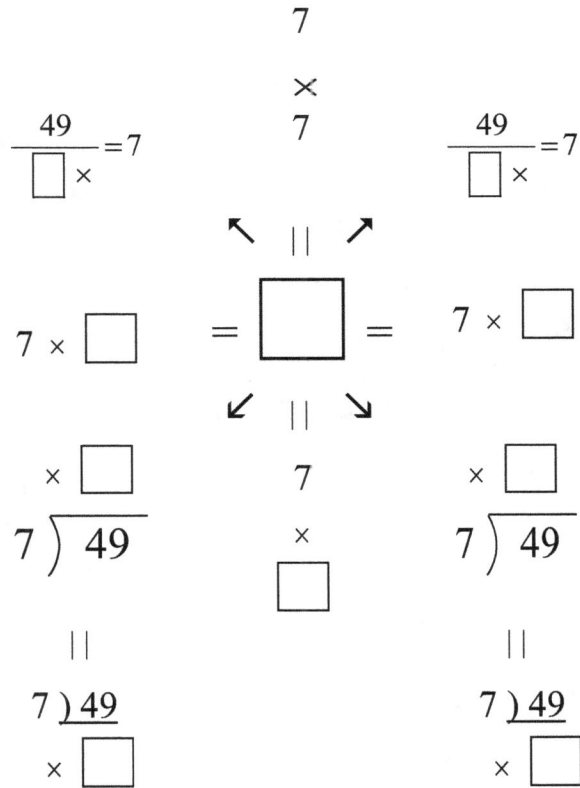

2007 - 2018 © Frank Ho, Amanda Ho, All rights reserved.   www.homathchess.com

Different ways of writing multiplication (Learning division while doing multiplications)

2007 - 2018 © Frank Ho, Amanda Ho, All rights reserved.   www.homathchess.com

Different ways of writing multiplication (Learning division while doing multiplications)

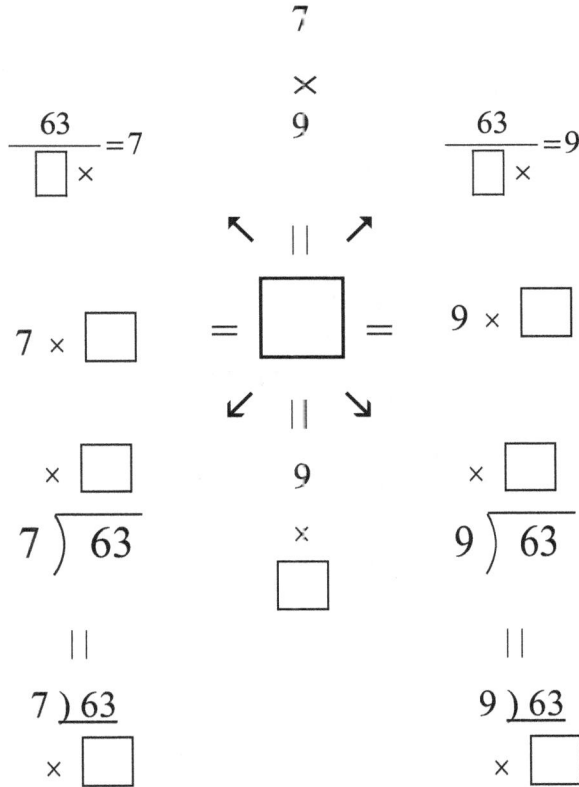

$$7 \times 9$$

$$\frac{63}{\boxed{} \times} = 7 \qquad \frac{63}{\boxed{} \times} = 9$$

$$7 \times \boxed{} = \boxed{} = 9 \times \boxed{}$$

$$\times \boxed{} \qquad 9 \qquad \times \boxed{}$$

$$7 \overline{)63} \qquad \times \qquad 9 \overline{)63}$$

$$\boxed{}$$

$$7 \underline{)63} \qquad\qquad 9 \underline{)63}$$

$$\times \boxed{} \qquad\qquad \times \boxed{}$$

2007 - 2018 © Frank Ho, Amanda Ho, All rights reserved.   www.homathchess.com

Counting 8's multiples

8, 16, 24, ☐,☐,☐,☐,☐,☐

Fill in the following ☐ with a number.

Sequence	♙	2	♗	4	5	6	7	8	♛
Add 8	☐	16	☐	32	☐	48	☐	64	☐

Sequence	1	2	♞	4	♜	6	7	8	9
Add 8	8	☐	24	☐	40	☐	56	☐	72

Sequence	♙	2	3	4	5	6	7	8	♛
Add 8	☐	16	☐	32	☐	48	☐	64	☐

Sequence	1	2	♗	4	♜	6	7	8	9
Add 8	8	☐	24	☐	40	☐	56	☐	72

Sequence	♙	2	3	4	5	6	7	8	♛
Add 8	☐	16	☐	32	☐	48	☐	64	☐

Sequence	1	2	♗	4	5	6	7	8	9
Add 8	8	☐	24	☐	40	☐	56	☐	72

2007 - 2018 © Frank Ho, Amanda Ho, All rights reserved.　　www.homathchess.com

8 times

8 × 1 = ☐	Eight times one is ☐	1 × 8 = ☐	One times eight is ☐
8 × 2 = ☐	Eight times two is ☐	2 × 8 = ☐	Two times eight is ☐
8 × 3 = ☐	Eight times three is ☐	♗ × 8 = ☐	Three times eight is ☐
8 × 4 = ☐	Eight times four is ☐	4 × 8 = ☐	Four times eight is ☐
8 × ♖ = ☐	Eight times five is ☐	5 × 8 = ☐	Five times eight is ☐
8 × 6 = ☐	Eight times six is ☐	6 × 8 = ☐	Six times eight is ☐
8 × 7 = ☐	Eight times seven is ☐	7 × 8 = ☐	Seven times eight is ☐
8 × 8 = ☐	Eight times eight is ☐	8 × 8 = ☐	Eight times eight is ☐
8 × ♕ = ☐	Eight times nine is ☐	9 × 8 = ☐	Nine times eight is ☐

Ho Math Chess 何数棋谜 妈！我会棋谜式乘法啦！

Mom! I Learn Multiplication Using Math-Chess-Puzzles Connection!

Student's Name _____ Date _____

2007 - 2018 © Frank Ho, Amanda Ho, All rights reserved. www.homathchess.com

8	1	8	2	8
X 1	X 8	X 2	X 8	X 3

8	6	7	8	♛
X 4	X 8	X 8	X 8	X 8

8	8	8	8	8
X ♜	X 8	X 7	X 8	X 6

5	8	7	4	9
X 8	X 6	X 8	X 8	X 8

7	♜	8	8	8
X 8	X 8	X 6	X 6	X 9

Student's Name _____ Date _____

2007 - 2018 © Frank Ho, Amanda Ho, All rights reserved.   www.homathchess.com

Oral practice

eight one eight	8 1 □	$\begin{array}{r} 1\,1 \\ \times\ \ 8 \\ \hline \square\,\square \end{array}$
eight two sixteen	8 2 □	$\begin{array}{r} 2\,2 \\ \times\ \ 8 \\ \hline \square\,\square\,\square \end{array}$
eight three twenty-four	8 ♞ □	$\begin{array}{r} 3\,3 \\ \times\ \ 8 \\ \hline \square\,\square\,\square \end{array}$
eight four thirty-two	8 4 □	$\begin{array}{r} 4\,4 \\ \times\ \ 8 \\ \hline \square\,\square\,\square \end{array}$
eight five forty	8 5 □	$\begin{array}{r} 4 \\ 5\,5 \\ \times\ \ 8 \\ \hline \square\,\square\,\square \end{array}$

2007 - 2018 © Frank Ho, Amanda Ho, All rights reserved.　www.homathchess.com

Oral practice

eight six forty-eight	8 6 ☐	$\overset{4}{6\,6} \times 8$ ☐☐☐
eight seven fifty-six	8 7 ☐	$7\,7 \times 8$ ☐☐☐
eight eight sixty-four	8 8 ☐	$8\,8 \times 8$ ☐☐☐
eight nine seventy-two	8 ♕ ☐	$9\,9 \times 8$ ☐☐☐
eight two sixteen	8 2 ☐	$2\,2 \times 8$ ☐☐☐
eight three twenty-four	8 ♘ ☐	$3\,3 \times 8$ ☐☐☐

Student's Name _____ Date _____

2007 - 2018 © Frank Ho, Amanda Ho, All rights reserved.   www.homathchess.com

Fill in ☐ with answer.

Times	Grouping	Addition
$8 \times \$1 = $ ☐	8 of ☐ = 8	$\$1 + \$1 + \$1 + \$1 + \$1 + \$1 + \$1 + \$1 = $ ☐
$1 \times \$8 = $ ☐	1 of ☐ = 8	8 of $\$1 = $ ☐

Fill in ☐ with answer.

Expression	Grouping	Addition
$8 \times \$2$	8 of ☐ = 16	$\$2 + \$2 + \$2 + \$2 + \$2 + \$2 + \$2 + \$2 = $ ☐
$2 \times \$8$	2 of ☐ = 16	$\$8 + \$8 = $ ☐

$8 \times 1 = $ ☐ $ = 1 \times 8 = $ ☐	$1 \times 8 = $ ☐ $ = 8 \times 1 = $ ☐
$8 \times $ ☐ $ = 16 = 2 \times $ ☐ $ = $ ☐	$2 \times $ ☐ $ = 16 = 8 \times $ ☐ $ = $ ☐

8 ♙ ☐	8 ♖ ☐	8 9 ☐	8 4 ☐	8 8 ☐
8 2 ☐	8 6 ☐	8 ♙ ☐	8 ♖ ☐	8 ♕ ☐
8 ♘ ☐	8 7 ☐	8 2 ☐	8 6 ☐	8 ♙ ☐
8 4 ☐	8 8 ☐	8 3 ☐	8 7 ☐	8 2 ☐

2007 - 2018 © Frank Ho, Amanda Ho, All rights reserved.　　www.homathchess.com

Fill in ☐ with answer.

Times	Grouping	Addition
8 × \$3 = ☐	8 of ☐ = 24	\$3 + \$3 + \$3 + \$3 + \$3 + \$3 + \$3 + \$3 = ☐
♞ × \$8 = ☐	3 of ☐ = 24	\$8 + \$8 + \$8 = ☐

Fill in ☐ with answer.

Times	Grouping	Addition
8 × \$4	8 of ☐ = 32	\$4 + \$4 + \$4 + \$4 + \$4 + \$4 + \$4 + \$4 = ☐
4 × \$8	4 of ☐ = 32	\$8 + \$8 + \$8 + \$8 = ☐

8 × 3 = ☐ = ♞ × 8 = ☐	3 × 8 = ☐ = 8 × 3 = ☐
8 × ☐ = 32 = 4 × ☐ = ☐	4 × ☐ = 32 = 8 × ☐ = ☐

8 1 ☐	8 ♜ ☐	8 9 ☐	8 4 ☐	8 8 ☐
8 2 ☐	8 6 ☐	8 1 ☐	8 5 ☐	8 9 ☐
8 ♞ ☐	8 7 ☐	8 2 ☐	8 6 ☐	8 ♟ ☐
8 4 ☐	8 8 ☐	8 ♞ ☐	8 7 ☐	8 2 ☐

2007 - 2018 © Frank Ho, Amanda Ho, All rights reserved.　www.homathchess.com

Fill in ☐ with answer.

Times	Grouping	Addition
8 × \$5 = ☐	8 of ☐ = 40	\$5 + \$5 + \$5 + \$5 + \$5 + \$5 + \$5 + \$5 = ☐
5 × \$8 = ☐	5 of ☐ = 40	S8 + \$8 + \$8 + \$8 + \$8 = ☐

Times	Grouping	Addition
8 × \$6 = ☐	8 of ☐ = 48	\$6 + \$6 + \$6 + \$6 + \$6 + \$6 + \$6 + \$6 = ☐
6 × \$8 = ☐	6 of ☐ = 48	\$8 + \$8 + \$8 + \$8 + \$8 + \$8 = ☐

8 × ♖ = ☐ = 5 × 8 = ☐	5 × 8 = ☐ = 8 × 5 = ☐
8 × ☐ = 48 = 6 × ☐ = ☐	6 × ☐ = 48 = 8 × ☐ = ☐

8 1 ☐	8 ♖ ☐	8 9 ☐	8 4 ☐	8 8 ☐
8 2 ☐	8 6 ☐	8 ♙ ☐	8 5 ☐	8 ♕ ☐
8 ♗ ☐	8 7 ☐	8 2 ☐	8 6 ☐	8 1 ☐
8 4 ☐	8 8 ☐	8 3 ☐	8 7 ☐	8 2 ☐

Student's Name _____ Date _____

2007 - 2018 © Frank Ho, Amanda Ho, All rights reserved.　www.homathchess.com

Fill in _____ and ☐ with answers.

Times	Grouping	Addition
8 × $7 = ☐	8 of ☐ = 56	$7 + $7 + $7 + $7 + $7 + $7 + $7 + $7 = ☐
7 × $8 = ☐	7 of ☐ = 56	$8 + $8 + $8 + $8 + $8 + $8 + $8 = ☐

Times	Grouping	Addition
8 × $8 = ☐	8 of ☐ = 64	$8 + $8 + $8 + $8 + $8 + $8 + $8 + $8 = ☐
8 × $8 = ☐	8 of ☐ = 64	$8 + $8 + $8 + $8 + $8 + $8 + $8 + $8 = ☐

8 × 7 = ☐ = 7 × 8 = ☐	7 × 8 = ☐ = 8 × 7 = ☐
8 × ☐ = 64 = 8 × ☐ = ☐	8 × ☐ = 64 = 8 × ☐ = ☐

8 1 ☐	8 5 ☐	8 ♛ ☐	8 4 ☐	8 8 ☐
8 2 ☐	8 6 ☐	8 1 ☐	8 ♜ ☐	8 9 ☐
8 ♞ ☐	8 7 ☐	8 2 ☐	8 6 ☐	8 ♙ ☐
8 4 ☐	8 8 ☐	8 3 ☐	8 7 ☐	8 2 ☐

2007 - 2018 © Frank Ho, Amanda Ho, All rights reserved.　　www.homathchess.com

Fill in ☐ with answer.

Times	Grouping	Addition
8 × $♛ = ☐	8 of ☐ = 72	$9 + $9 + $9 + $9 + $9 + $9 + $9 + $9 = ☐
9 × $8 = ☐	9 of ☐ = 72	$8 + $8 + $8 + $8 + $8 + $8 + $8 + $8+ $8= ☐

Times	Grouping	Addition
8 × $8 = ☐	8 of ☐ = 64	$8 + $8 + $8 + $8 + $8 + $8 + $8 + $8 = ☐
8 × $8 = ☐	8 of ☐ = 64	$8 + $8 + $8 + $8 + $8 + $8 + $8 + $8 = ☐

8 × 9 = ☐ = 9 × 8 = ☐	9 × 8 = ☐ = 8 × 9 = ☐
8 × ☐ = 64 = 8 × ☐ = ☐	8 × ☐ = 64 = 8 × ☐ = ☐

8 ♙ ☐	8 ♖ ☐	8 ♕ ☐	8 4 ☐	8 8 ☐
8 2 ☐	8 6 ☐	8 1 ☐	8 ♜ ☐	8 9 ☐
8 ♝ ☐	8 7 ☐	8 2 ☐	8 6 ☐	8 ♙ ☐
8 4 ☐	8 8 ☐	8 3 ☐	8 7 ☐	8 2 ☐

Ho Math Chess 何数棋谜 妈!我会棋谜式乘法啦!
Mom! I Learn Multiplication Using Math-Chess-Puzzles Connection!

Student's Name _____ Date _____

2007 - 2018 © Frank Ho, Amanda Ho, All rights reserved.   www.homathchess.com

Preparing for division

□	□	□	□	□
X 5	X 2	X 4	X 9	X ♗
40	16	32	72	24
□	□	□	□	□
X 7	X 6	X 8	X ♗	X 8
56	48	24	6	48
□	□	□	□	□
X 4	X 8	X 5	X 2	X 7
32	72	40	16	56
□	□	□	□	□
X 8	X 8	X 8	X 3	X 8
56	48	32	24	16
□	□	□	□	□
X 4	X ♖	X 4	X 2	X 8
32	40	32	16	56

Ho Math Chess　何数棋谜　妈！我会棋谜式乘法啦！
Mom! I Learn Multiplication Using Math-Chess-Puzzles Connection!

Student's Name _____ Date _____

2007 - 2018 © Frank Ho, Amanda Ho, All rights reserved.　www.homathchess.com

Preparing for division

☐ X 8 = 8	X ☐ 8)8̄	☐)8 X 8
☐ X 8 = 16	X ☐ 8)16	☐)16 X 8
☐ X 8 = 24	X ☐ 8)24	☐)24 X 8
☐ X 8 = 32	X ☐ 8)32	☐)32 X 8
☐ X 8 = 40	X ☐ 8)40	☐)40 X 8
☐ X 8 = 48	X ☐ 8)48	☐)48 X 8
☐ X 8 = 56	X ☐ 8)56	☐)56 X 8

Mom! I Learn Multiplication Using Math-Chess-Puzzles Connection!

Student's Name _____ Date _____

2007 - 2018 © Frank Ho, Amanda Ho, All rights reserved.　www.homathchess.com

Preparing for division

☐ X 8 = 64	X ☐ 8) 64	☐) 64 X 8
☐ X 8 = 72	X ☐ 8) 72	☐) 72 X 8
☐ X 8 = 16	X ☐ 8) 16	☐) 16 X 8
☐ X 8 = 24	X ☐ 8) 24	☐) 24 X 8
☐ X 8 = 32	X ☐ 8) 32	☐) 32 X 8
☐ X 8 = 40	X ☐ 8) 40	☐) 40 X 8
☐ X 8 = 48	X ☐ 8) 48	☐) 48 X 8

Student's Name _____ Date _____

2007 - 2018 © Frank Ho, Amanda Ho, All rights reserved.   www.homathchess.com

Cross multiplication

12 12 ↖ ↗ $\frac{6}{2} = \frac{6}{2}$	☐ ☐ ↖ ↗ $\frac{8}{8} = \frac{2}{2}$	☐ ☐ ↖ ↗ $\frac{8}{8} = \frac{3}{3}$	☐ ☐ ↖ ↗ $\frac{8}{8} = \frac{4}{4}$
☐ ☐ ↖ ↗ $\frac{8}{8} = \frac{5}{5}$	☐ ☐ ↖ ↗ $\frac{8}{8} = \frac{6}{6}$	☐ ☐ ↖ ↗ $\frac{8}{8} = \frac{7}{7}$	☐ ☐ ↖ ↗ $\frac{8}{8} = \frac{9}{9}$
☐ ☐ ↖ ↗ $\frac{8}{8} = \frac{5}{5}$	☐ ☐ ↖ ↗ $\frac{8}{8} = \frac{4}{4}$	☐ ☐ ↖ ↗ $\frac{8}{8} = \frac{7}{7}$	☐ ☐ ↖ ↗ $\frac{8}{8} = \frac{8}{8}$
☐ ☐ ↖ ↗ $\frac{8}{8} = \frac{6}{6}$	☐ ☐ ↖ ↗ $\frac{8}{8} = \frac{9}{9}$	☐ ☐ ↖ ↗ $\frac{8}{8} = \frac{8}{8}$	☐ ☐ ↖ ↗ $\frac{8}{8} = \frac{3}{3}$

2007 - 2018 © Frank Ho, Amanda Ho, All rights reserved.   www.homathchess.com

Different ways of writing multiplication (Learning division while doing multiplications)

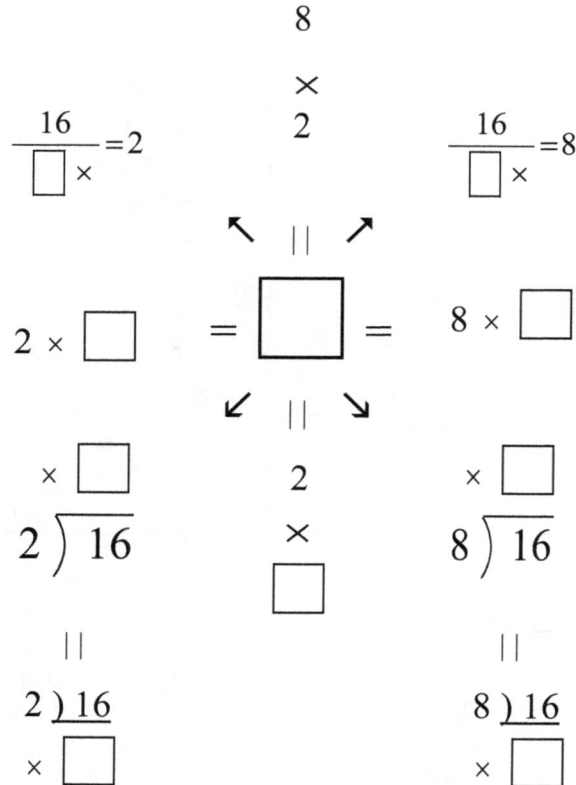

Different ways of writing multiplication (Learning division while doing multiplications)

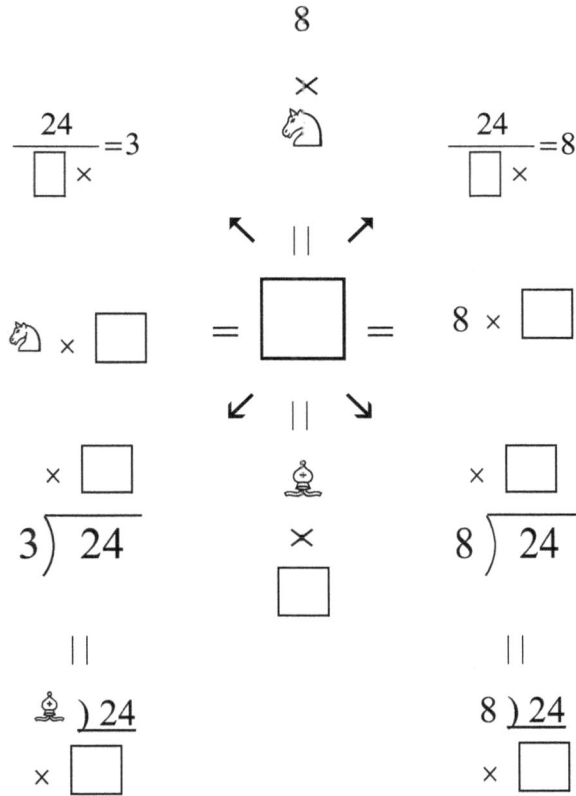

2007 - 2018 © Frank Ho, Amanda Ho, All rights reserved.　　www.homathchess.com

Different ways of writing multiplication (Learning division while doing multiplications)

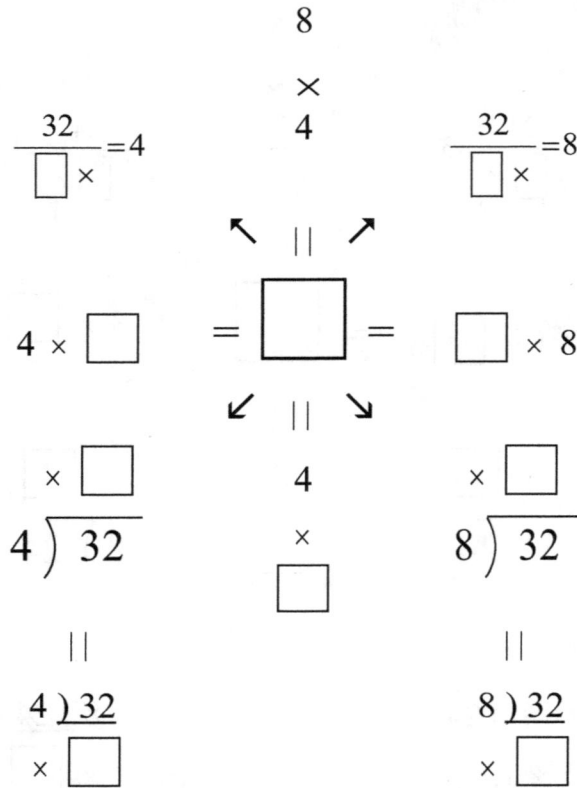

2007 - 2018 © Frank Ho, Amanda Ho, All rights reserved.　　www.homathchess.com

Different ways of writing multiplication (Learning division while doing multiplications)

2007 - 2018 © Frank Ho, Amanda Ho, All rights reserved.　　www.homathchess.com

Different ways of writing multiplication (Learning division while doing multiplications)

$$8 \times 6$$

$$\frac{48}{\square \times} = 6 \qquad \frac{48}{\square \times} = 8$$

$$6 \times \square = \boxed{} = 8 \times \square$$

$$\times \square$$
$$6 \overline{)\,48}$$

$$\begin{array}{c} 6 \\ \times \\ \square \end{array}$$

$$\times \square$$
$$8 \overline{)\,48}$$

$$6 \underline{)\,48}$$
$$\times \square$$

$$8 \underline{)\,48}$$
$$\times \square$$

2007 - 2018 © Frank Ho, Amanda Ho, All rights reserved.   www.homathchess.com

Different ways of writing multiplication (Learning division while doing multiplications)

$$\frac{56}{\Box \times} = 7 \qquad \begin{array}{c} 8 \\ \times \\ 7 \end{array} \qquad \frac{56}{\Box \times} = 8$$

$$7 \times \Box \quad = \quad \boxed{} \quad = \quad 8 \times \Box$$

$$7\,\overline{)\,56} \qquad \begin{array}{c} 7 \\ \times \\ \Box \end{array} \qquad 8\,\overline{)\,56}$$

$$7\,\overline{)\,56} \qquad\qquad 8\,\overline{)\,56}$$
$$\times \Box \qquad\qquad\qquad \times \Box$$

2007 - 2018 © Frank Ho, Amanda Ho, All rights reserved.   www.homathchess.com

Different ways of writing multiplication (Learning division while doing multiplications)

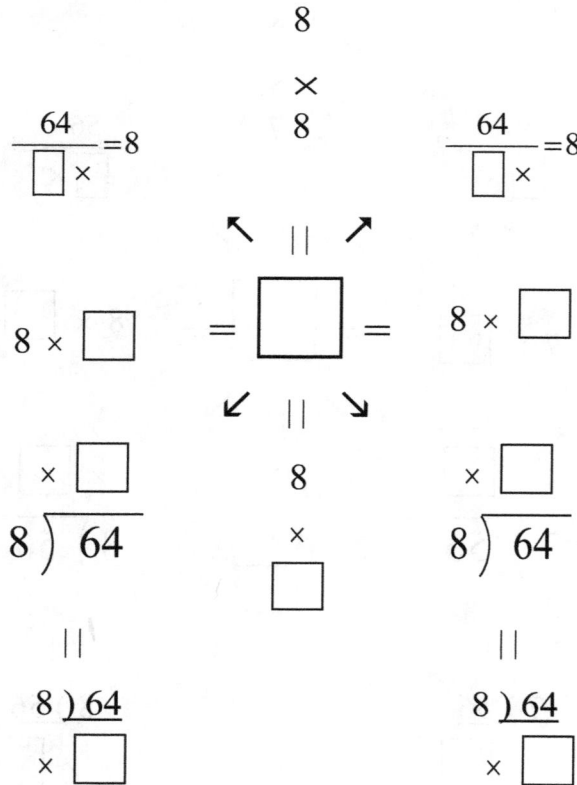

$$8 \times 8$$

$$\frac{64}{\boxed{} \times} = 8 \qquad\qquad \frac{64}{\boxed{} \times} = 8$$

$$8 \times \boxed{} = \boxed{} = 8 \times \boxed{}$$

$$\times \boxed{} \qquad\qquad 8 \qquad\qquad \times \boxed{}$$

$$8\,\overline{)\,64} \qquad \times \qquad 8\,\overline{)\,64}$$

$$\boxed{}$$

$$8\,)\,64 \qquad\qquad 8\,)\,64$$

$$\times \boxed{} \qquad\qquad \times \boxed{}$$

2007 - 2018 © Frank Ho, Amanda Ho, All rights reserved.   www.homathchess.com

Different ways of writing multiplication (Learning division while doing multiplications)

$$8 \times \text{♛}$$

$$\frac{72}{\Box \times} = 8 \qquad \frac{72}{\Box \times} = 9$$

$$8 \times \Box \qquad = \boxed{} = \qquad 9 \times \Box$$

$$\times \Box \qquad \qquad \times \Box$$

$$8 \overline{)\,72} \qquad 9 \times \qquad 9 \overline{)\,72}$$

$$\Box$$

$$8 \,\overline{)\,72} \qquad\qquad \text{♛} \,\overline{)\,72}$$
$$\times \Box \qquad\qquad \times \Box$$

Ho Math Chess　何数棋谜　妈！我会棋谜式乘法啦！

Mom! I Learn Multiplication Using Math-Chess-Puzzles Connection!

Student's Name _____ Date _____

2007 - 2018 © Frank Ho, Amanda Ho, All rights reserved.　　www.homathchess.com

Counting 9s multiples

9, 18, 27, □,□,□,□,□,□

Fill in the following □ with a number.

Sequence	1	2	♞	4	♜	6	7	8	♛
Add 9	□	18	□	36	□	54	□	72	□

Sequence	♙	2	3	4	5	6	7	8	9
Add 9	9	□	27	□	45	□	63	□	81

Sequence	1	2	♞	4	♜	6	7	8	♛
Add 9	□	18	□	36	□	54	□	72	□

Sequence	1	2	3	4	5	6	7	8	9
Add 9	9	□	27	□	45	□	63	□	81

Sequence	♙	2	♞	4	♜	6	7	8	♛
Add 9	□	18	□	36	□	54	□	72	□

Sequence	1	2	3	4	♜	6	7	8	9
Add 9	9	□	27	□	45	□	63	□	81

Student's Name _____ Date _____

2007 - 2018 © Frank Ho, Amanda Ho, All rights reserved.   www.homathchess.com

9 times

9 × 1 = ☐	Nine times one is ☐	1 × 9 = ☐	One times nine is ☐
9 × 2 = ☐	Nine times two is ☐	2 × 9 = ☐	Two times nine is ☐
9 × 3 = ☐	Nine times three is ☐	3 × 9 = ☐	Three times nine is ☐
9 × 4 = ☐	Nine times four is ☐	4 × 9 = ☐	Four times nine is ☐
9 × 5 = ☐	Nine times five is ☐	5 × 9= ☐	Five times nine is ☐
9 × 6 = ☐	Nine times six is ☐	6 × 9 = ☐	Six times nine is ☐
9 × 7 = ☐	Nine times seven is ☐	7 × 9 = ☐	Seven times nine is ☐
9 × 8 = ☐	Nine times eight is ☐	8 × 9 = ☐	Eight times nine is ☐
9 × 9 = ☐	Nine times nine is ☐	9 × 9 = ☐	Nine times nine is ☐

```
    9          1          2          ♛          3
  X 1        X 9        X 9        X 2        X 9
   ☐          ☐        ☐☐        ☐☐        ☐☐

    5          9          9          8          9
  X ♛        X 6        X 7        X ♛        X 3
  ☐☐        ☐☐        ☐☐        ☐☐        ☐☐

    4          3          5          7          9
  X 9        X ♛        X 9        X 9        X ♛
  ☐☐        ☐☐        ☐☐        ☐☐        ☐☐
```

2007 - 2018 © Frank Ho, Amanda Ho, All rights reserved.　www.homathchess.com

9	1	9	2	9
X 1	X ♛	X 2	X 9	X 3

9	9	7	♛	9
X 4	X 5	X 9	X 8	X 6

♛	6	9	6	9
X 6	X 9	X 7	X ♛	X 8

9	6	7	9	9
X 5	X ♛	X 9	X 8	X 4

7	9	5	9	6
X 9	X 8	X 9	X ♗	X 9

2007 - 2018 © Frank Ho, Amanda Ho, All rights reserved.   www.homathchess.com

Oral practice

nine one nine	9 1 ☐	$\begin{array}{r} 1\,1 \\ \times\ \ 9 \\ \hline \square\square \end{array}$
nine two eighteen	♛ 2 ☐	$\begin{array}{r} 2\,2 \\ \times\ \ 9 \\ \hline \square\square\square \end{array}$
nine three twenty-seven	9 ♝ ☐	$\begin{array}{r} 3\,3 \\ \times\ \ 9 \\ \hline \square\square\square \end{array}$
nine four thirty-six	9 4 ☐	$\begin{array}{r} 4\,4 \\ \times\ \ ♛ \\ \hline \square\square\square \end{array}$
nine five forty-five	♛ ♜ ☐	$\begin{array}{r} {}^{4}\ \ \\ 5\,5 \\ \times\ \ 9 \\ \hline \square\square\square \end{array}$

Student's Name _____ Date _____

2007 - 2018 © Frank Ho, Amanda Ho, All rights reserved.   www.homathchess.com

Oral practice

nine six fifty-four	9 6 ☐	¹ 6 6 × ♛ ☐ ☐ ☐
nine seven sixty-three	♛ 7 ☐	7 7 × 9 ☐ ☐ ☐
nine eight seventy-two	9 8 ☐	8 8 × 9 ☐ ☐ ☐
nine nine eighty-one	9 ♛ ☐	9 9 × 9 ☐ ☐ ☐
nine four thirty-six	9 4 ☐	4 4 × 9 ☐ ☐ ☐

Student's Name _____ Date _____

2007 - 2018 © Frank Ho, Amanda Ho, All rights reserved.   www.homathchess.com

Preparing for division

□	□	□	□	□
X 2	X ♗	X 4	X 5	X 6
18	27	36	45	54

□	□	□	□	□
X 7	X 8	X 9	X 3	X 4
63	72	81	27	36

□	□	□	□	□
X 9	X 9	X 5	X 2	X 7
45	18	45	18	63

□	□	□	□	□
X 4	X ♖	X 4	X ♗	X 8
36	45	36	27	72

□	□	□	□	□
X 4	X 5	X 6	X 2	X 7
36	45	54	18	63

Ho Math Chess 何数棋谜 妈!我会棋谜式乘法啦!

Mom! I Learn Multiplication Using Math-Chess-Puzzles Connection!

Student's Name _____ Date _____

2007 - 2018 © Frank Ho, Amanda Ho, All rights reserved.   www.homathchess.com

Fill in ☐ with answer.

Times	Grouping	Addition
$9 \times \$1 = $ ☐	9 of ☐ = 9	$\$1 + \$1 + \$1 + \$1 + \$1 + \$1 + \$1 + \$1 + \$1 = $ ☐
$1 \times \$9 = $ ☐	1 of ☐ = 9	9 of $\$1 = $ ☐

Fill in ☐ with answer.

Expression	Grouping	Addition
$9 \times \$2$	9 of ☐ = 18	$\$2 + \$2 + \$2 + \$2 + \$2 + \$2 + \$2 + \$2 + \$2 = $ ☐
$2 \times \$9$	2 of ☐ = 18	$\$9 + \$9 = $ ☐

$9 \times 1 = $ ☐ $ = ♙ \times 9 = $ ☐	$1 \times 9 = $ ☐ $ = 9 \times 1 = $ ☐
$9 \times $ ☐ $ = 18 = 2 \times $ ☐ $ = $ ☐	$2 \times $ ☐ $ = 18 = 9 \times $ ☐ $ = $ ☐

9 ♙ ☐	9 5 ☐	9 9 ☐	9 4 ☐	9 8 ☐
9 2 ☐	♛ 6 ☐	9 1 ☐	9 ♜ ☐	♛ 9 ☐
♛ 3 ☐	9 7 ☐	♛ 2 ☐	9 6 ☐	9 1 ☐
9 4 ☐	9 8 ☐	9 3 ☐	9 7 ☐	♛ 2 ☐

Ho Math Chess 何数棋谜 妈!我会棋谜式乘法啦!
Mom! I Learn Multiplication Using Math-Chess-Puzzles Connection!

Student's Name _____ Date _____

2007 - 2018 © Frank Ho, Amanda Ho, All rights reserved.   www.homathchess.com

Fill in ☐ with answer.

Times	Grouping	Addition
9 × \$3 = ☐	9 of ☐ = 27	\$3 + \$3 + \$3 + \$3 + \$3 + \$3 + \$3 + \$3 + \$3 = ☐
3 × \$9 = ☐	3 of ☐ = 27	\$9 + \$9 + \$9 = ☐

Fill in ☐ with answer.

Times	Grouping	Addition
9 × \$4	9 of ☐ = 36	\$4 + \$4 + \$4 + \$4 + \$4 + \$4 + \$4 + \$4 + \$4 = ☐
4 × \$9	4 of ☐ = 36	\$9 + \$9 + \$9 + \$9 = ☐

9 × 3 = ☐ = 3 × 9 = ☐	3 × 9 = ☐ = 9 × 3 = ☐
9 × ☐ = 36 = 4 × ☐ = ☐	4 × ☐ = 36 = 9 × ☐ = ☐

♛ 1 ☐	9 ♜ ☐	♛ 9 ☐	9 4 ☐	9 8 ☐
9 2 ☐	9 6 ☐	9 1 ☐	9 5 ☐	9 ♛ ☐
9 ♗ ☐	♛ 7 ☐	9 2 ☐	♛ 6 ☐	9 1 ☐
9 4 ☐	9 8 ☐	9 3 ☐	9 7 ☐	9 2 ☐

2007 - 2018 © Frank Ho, Amanda Ho, All rights reserved.   www.homathchess.com

Fill in ☐ with answer.

Times	Grouping	Addition
$9 \times \$5 = \square$	9 of $\square$ = 45	$\$5 + \$5 + \$5 + \$5 + \$5 + \$5 + \$5 + \$5 + \$5 = \square$
$5 \times \$9 = \square$	5 of $\square$ = 45	$\$9 + \$9 + \$9 + \$9 + \$9 = \square$

Times	Grouping	Addition
$9 \times \$6 = \square$	9 of $\square$ = 54	$\$6 + \$6 + \$6 + \$6 + \$6 + \$6 + \$6 + \$6 + \$6 = \square$
$6 \times \$9 = \square$	6 of $\square$ = 54	$\$9 + \$9 + \$9 + \$9 + \$9 + \$9 = \square$

$9 \times 5 = \square = ♖ \times 9 = \square$	$5 \times 9 = \square = 9 \times 5 = \square$
$9 \times \square = 54 = 6 \times \square = \square$	$6 \times \square = 54 = 9 \times \square = \square$

♕ 1 $\square$	9 5 $\square$	♕ 9 $\square$	9 4 $\square$	9 8 $\square$
9 2 $\square$	9 6 $\square$	9 ♙ $\square$	9 ♖ $\square$	9 9 $\square$
9 3 $\square$	♕ 7 $\square$	9 2 $\square$	♕ 6 $\square$	9 ♙ $\square$
9 4 $\square$	9 8 $\square$	9 3 $\square$	9 7 $\square$	♕ 2 $\square$

Student's Name _____ Date _____

2007 - 2018 © Frank Ho, Amanda Ho, All rights reserved.   www.homathchess.com

Fill in _____ and ☐ with answers.

Times	Grouping	Addition
9 × $7 = ☐	9 of ☐ = 63	$7 + $7 + $7 + $7 + $7 + $7 + $7 + $7 + $7 = ☐
7 × $9 = ☐	7 of ☐ = 63	$9 + $9 + $9 + $9 + $9 + $9 + $9 = ☐

Times	Grouping	Addition
9 × $8 = ☐	9 of ☐ = 72	$8 + $8 + $8 + $8 + $8 + $8 + $8 + $8 + $8 = ☐
8 × $9 = ☐	8 of ☐ = 72	$9 + $9 + $9 + $9 + $9 + $9 + $9 + $9 = ☐

9 × 7 = ☐ = 7 × 9 = ☐	7 × 9 = ☐ = 9 × 7 = ☐
9 × ☐ = 72 = 8 × ☐ = ☐	8 × ☐ = 72 = 9 × ☐ = ☐

9 1 ☐	9 5 ☐	9 ♛ ☐	9 4 ☐	♛ 8 ☐
♛ 2 ☐	9 6 ☐	9 1 ☐	9 ♜ ☐	9 9 ☐
9 3 ☐	9 7 ☐	9 2 ☐	♛ 6 ☐	9 1 ☐
9 4 ☐	♛ 8 ☐	9 ♝ ☐	9 7 ☐	♛ 2 ☐

Student's Name _____ Date _____

2007 - 2018 © Frank Ho, Amanda Ho, All rights reserved.   www.homathchess.com

Fill in ☐ with answer.

Times	Grouping	Addition
9 × $9 = ☐	9 of ☐ = 72	$9 + $9 + $9 + $9 + $9 + $9 + $9 + $9 + $9 = ☐
9 × $9 = ☐	8 of ☐ = 72	$9 + $9 + $9 + $9 + $9 + $9 + $9 + $9 + $9 = ☐

Times	Grouping	Addition
9 × $8 = ☐	9 of ☐ = 72	$8 + $8 + $8 + $8 + $8 + $8 + $8 + $8 + $8 = ☐
8 × $9 = ☐	8 of ☐ = 72	$9 + $9 + $9 + $9 + $9 + $9 + $9 + $9 = ☐

9 × 9 = ☐ = 9 × ♛ = ☐	9 × 9 = ☐ = ♛ × 9 = ☐
9 × ☐ = 81 = 9 × ☐ = ☐	9 × ☐ = 81 = 9 × ☐ = ☐

9 1 ☐	9 5 ☐	♛ 9 ☐	9 4 ☐	9 8 ☐
9 2 ☐	♛ 6 ☐	9 1 ☐	9 ♜ ☐	9 ♛ ☐
9 3 ☐	9 7 ☐	9 2 ☐	♛ 6 ☐	9 1 ☐
♛ 4 ☐	9 8 ☐	9 ♗ ☐	9 7 ☐	9 2 ☐

2007 - 2018 © Frank Ho, Amanda Ho, All rights reserved.   www.homathchess.com

Preparing for division

□ X 9 = 9	X □ 9)9	□)9 X 9
□ X ♛ = 18	X □ 9)18	□)18 X ♛
□ X 9 = 27	X □ 9)27	□)27 X 9
□ X ♛ = 36	X □ 9)36	□)36 X ♛
□ X 9 = 45	X □ 9)45	□)45 X 9
□ X ♛ = 54	X □ 9)54	□)54 X ♛
□ X 9 = 63	X □ 9)63	□)63 X 9

2007 - 2018 © Frank Ho, Amanda Ho, All rights reserved.   www.homathchess.com

Preparing for division

☐ X 9 = 72	X ☐ 9)72	☐) 72 X ♛
☐ X 9 = 81	X ☐ 9)81	☐) 81 X 9
☐ X ♛ = 9	X ☐ 9)9	☐) 9 X 9
☐ X 9 = 18	X ☐ 9)18	☐) 18 X ♛
☐ X 9 = 27	X ☐ 9)27	☐) 27 X 9
☐ X ♛ = 36	X ☐ 9)36	☐) 36 X 9
☐ X 9 = 45	X ☐ 9)45	☐) 45 X ♛

2007 - 2018 © Frank Ho, Amanda Ho, All rights reserved.   www.homathchess.com

Cross multiplication

12 12 ↖ ↗ $\dfrac{6}{2} = \dfrac{6}{2}$	▢ ▢ ↖ ↗ $\dfrac{9}{9} = \dfrac{2}{2}$	▢ ▢ ↖ ↗ $\dfrac{9}{9} = \dfrac{3}{3}$	▢ ▢ ↖ ↗ $\dfrac{9}{9} = \dfrac{4}{4}$
▢ ▢ ↖ ↗ $\dfrac{9}{9} = \dfrac{5}{5}$	▢ ▢ ↖ ↗ $\dfrac{9}{9} = \dfrac{6}{6}$	▢ ▢ ↖ ↗ $\dfrac{9}{9} = \dfrac{7}{7}$	▢ ▢ ↖ ↗ $\dfrac{9}{9} = \dfrac{9}{9}$
▢ ▢ ↖ ↗ $\dfrac{9}{9} = \dfrac{5}{5}$	▢ ▢ ↖ ↗ $\dfrac{9}{9} = \dfrac{4}{4}$	▢ ▢ ↖ ↗ $\dfrac{9}{9} = \dfrac{7}{7}$	▢ ▢ ↖ ↗ $\dfrac{9}{9} = \dfrac{8}{8}$
▢ ▢ ↖ ↗ $\dfrac{9}{9} = \dfrac{6}{6}$	▢ ▢ ↖ ↗ $\dfrac{9}{9} = \dfrac{8}{8}$	▢ ▢ ↖ ↗ $\dfrac{9}{9} = \dfrac{9}{9}$	▢ ▢ ↖ ↗ $\dfrac{9}{9} = \dfrac{3}{3}$

2007 - 2018 © Frank Ho, Amanda Ho, All rights reserved.　　www.homathchess.com

Different ways of writing multiplication (Learning division while doing multiplications)

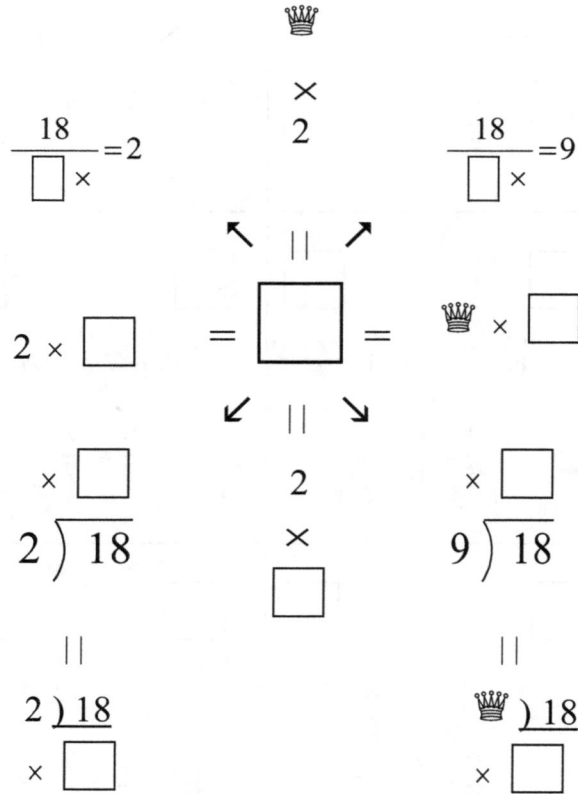

$$\frac{18}{\Box \times} = 2 \qquad \qquad \frac{18}{\Box \times} = 9$$

$$2 \times \Box \qquad = \Box = \text{♕} \times \Box$$

$$2\overline{)18} \qquad \qquad 9\overline{)18}$$

$$2\,)\,18 \qquad \qquad \text{♕}\,)\,18$$

Ho Math Chess   何数棋谜   妈！我会棋谜式乘法啦！
Mom! I Learn Multiplication Using Math-Chess-Puzzles Connection!

Student's Name _____ Date _____

2007 - 2018 © Frank Ho, Amanda Ho, All rights reserved.   www.homathchess.com

Different ways of writing multiplication (Learning division while doing multiplications)

$$9$$
$$\times$$
♗

$$\frac{27}{\square \times} = 3$$
$$\frac{27}{\square \times} = 9$$

↖ ‖ ↗

♗ $\times \square$ $=$ $\boxed{}$ $=$ ♕ $\times \square$

↙ ‖ ↘

$\times \square$
$$3 \overline{)\, 27}$$

$$3$$
$$\times$$
$$\square$$

$\times \square$
$$9 \overline{)\, 27}$$

‖

♘ $)\, 27$
$\times \square$

‖

♕ $)\, 27$
$\times \square$

Mom! I Learn Multiplication Using Math-Chess-Puzzles Connection!

Student's Name _____ Date _____

2007 - 2018 © Frank Ho, Amanda Ho, All rights reserved.   www.homathchess.com

Different ways of writing multiplication (Learning division while doing multiplications)

Different ways of writing multiplication (Learning division while doing multiplications)

$$9 \times ♖$$

$$\frac{45}{\Box \times} = 5 \qquad\qquad \frac{45}{\Box \times} = 9$$

$$↖ \ || \ ↗$$

$$♖ \times \Box = \boxed{} = 9 \times \Box$$

$$↙ \ || \ ↘$$

$$\times \Box \qquad\qquad 5 \qquad\qquad \times \Box$$

$$5 \overline{)\,45} \qquad \times \Box \qquad 9 \overline{)\,45}$$

$$||\qquad\qquad\qquad\qquad ||$$

$$5\,\overline{)\,45} \qquad\qquad ♕\,\overline{)\,45}$$

$$\times \Box \qquad\qquad\qquad \times \Box$$

2007 - 2018 © Frank Ho, Amanda Ho, All rights reserved.   www.homathchess.com

Different ways of writing multiplication (Learning division while doing multiplications)

$$9 \times 6$$

$$\frac{54}{\boxed{}\times} = 6 \qquad \frac{54}{\boxed{}\times} = 9$$

$$6 \times \boxed{} \quad = \boxed{} = \text{♛} \times \boxed{}$$

$$\times \boxed{} \qquad\qquad \times \boxed{}$$

$$6\overline{)54} \qquad 6 \qquad 9\overline{)54}$$

$$\times \boxed{}$$

$$6\,)\,54 \qquad\qquad \text{♛}\,)\,54$$

$$\times \boxed{} \qquad\qquad \times \boxed{}$$

2007 - 2018 © Frank Ho, Amanda Ho, All rights reserved.   www.homathchess.com

Different ways of writing multiplication (Learning division while doing multiplications)

$$9 \times 7$$

$$\frac{63}{\square \times} = 7 \qquad \frac{63}{\square \times} = 9$$

$$7 \times \square \;=\; \boxed{} \;=\; ♛ \times \square$$

$$7) \overline{63} \qquad 7 \times \square \qquad 9) \overline{63}$$

$$7 \,\underline{)\,63} \qquad\qquad ♛ \,\underline{)\,63}$$
$$\times \square \qquad\qquad\qquad \times \square$$

2007 - 2018 © Frank Ho, Amanda Ho, All rights reserved.　　www.homathchess.com

Different ways of writing multiplication (Learning division while doing multiplications)

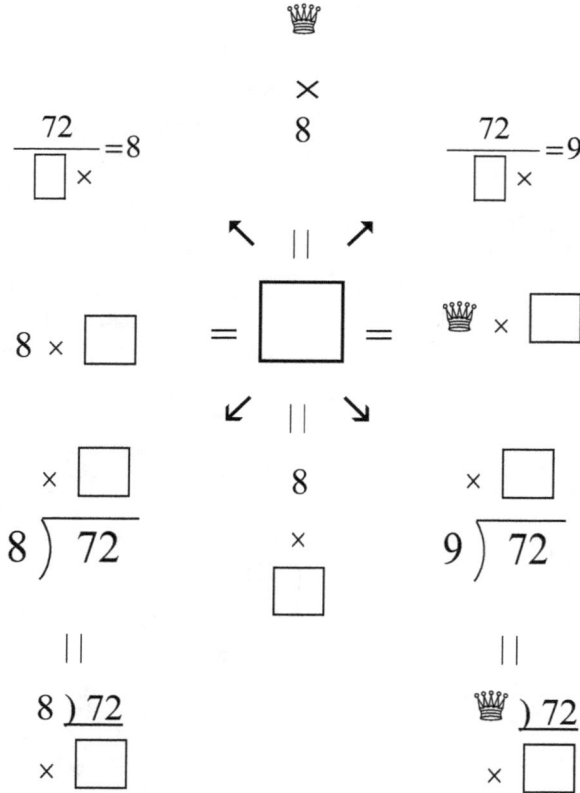

2007 - 2018 © Frank Ho, Amanda Ho, All rights reserved.   www.homathchess.com

Different ways of writing multiplication (Learning division while doing multiplications)

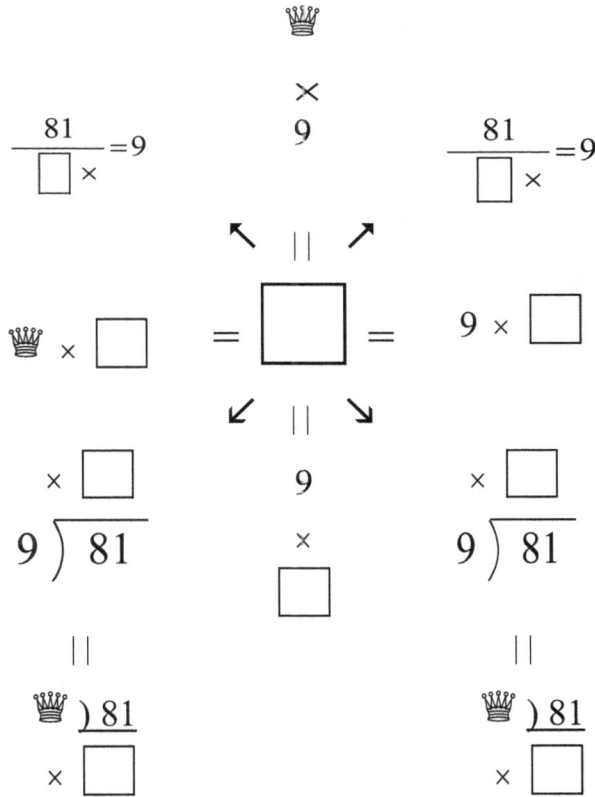

Ho Math Chess　何数棋谜　妈！我会棋谜式乘法啦！
Mom! I Learn Multiplication Using Math-Chess-Puzzles Connection!

Student's Name _____ Date _____

2007 - 2018 © Frank Ho, Amanda Ho, All rights reserved.　www.homathchess.com

Product using image process

White has _____ product points.

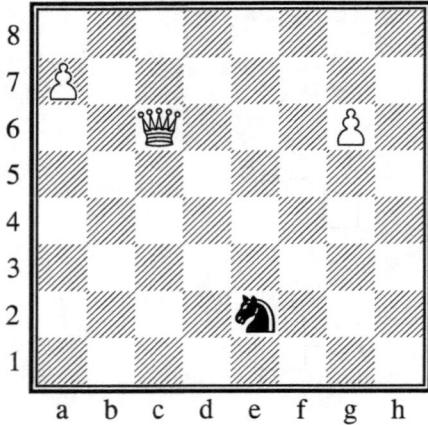

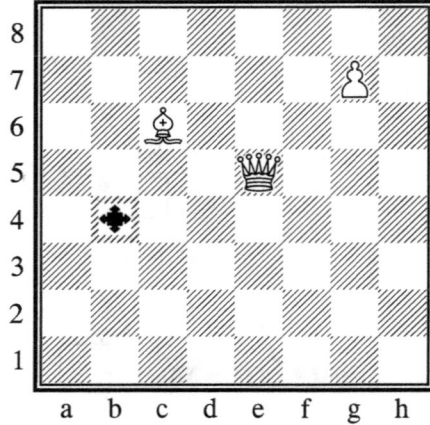

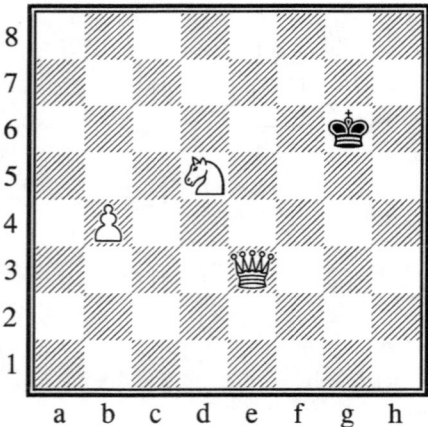

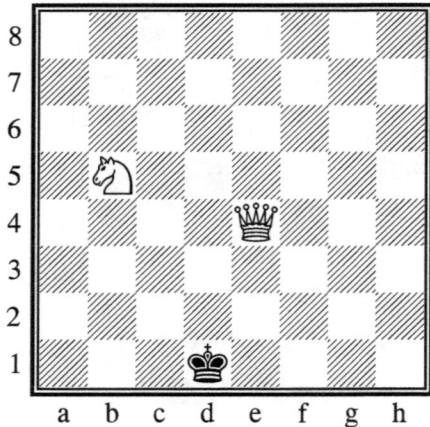

White has _____ product points.

White has _____ product points.

White has _____ product points.

196

Mom! I Learn Multiplication Using Math-Chess-Puzzles Connection!

Student's Name _____ Date _____

2007 - 2018 © Frank Ho, Amanda Ho, All rights reserved.   www.homathchess.com

Product using image process

White has _____ product points.

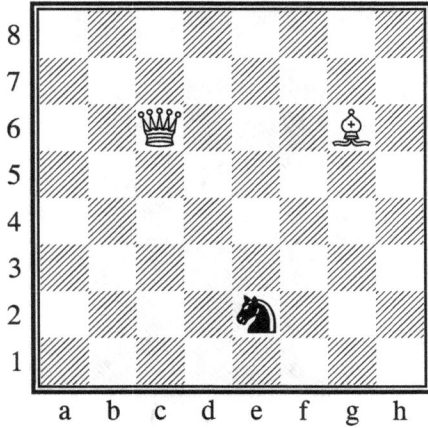

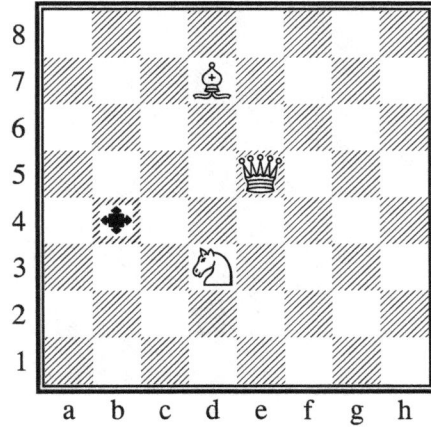

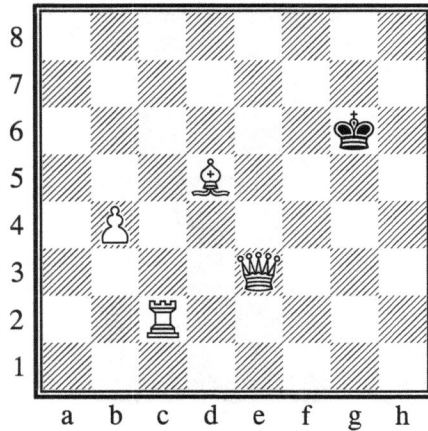

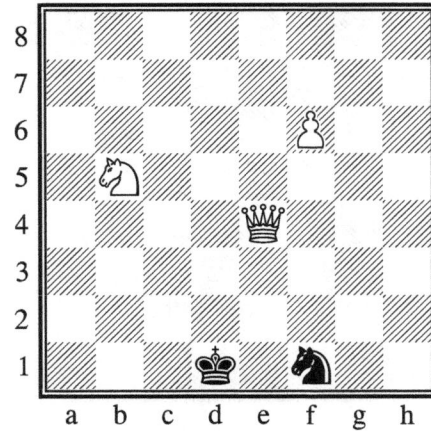

White has _____ product points.

White has _____ product points.

White has _____ product points.

Ho Math Chess   何数棋谜   妈！我会棋谜式乘法啦！
Mom! I Learn Multiplication Using Math-Chess-Puzzles Connection!

Student's Name _____ Date _____

2007 - 2018 © Frank Ho, Amanda Ho, All rights reserved.   www.homathchess.com

Product using image process

White has _____ product points.

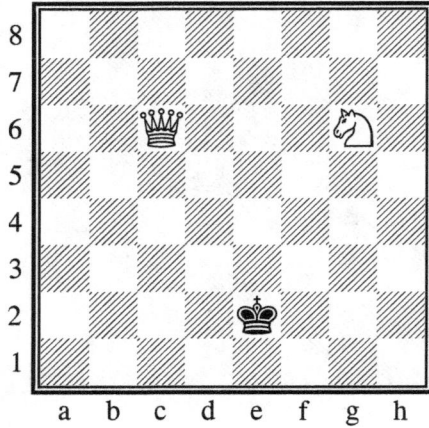

White has _____ product points.

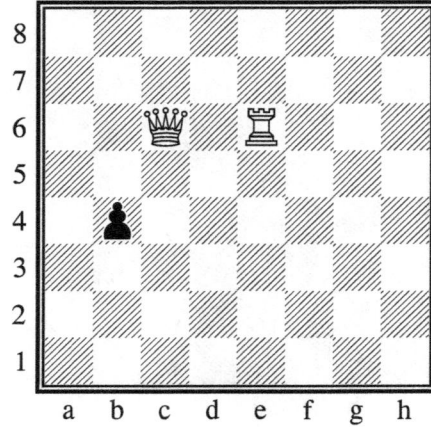

White has _____ product points.

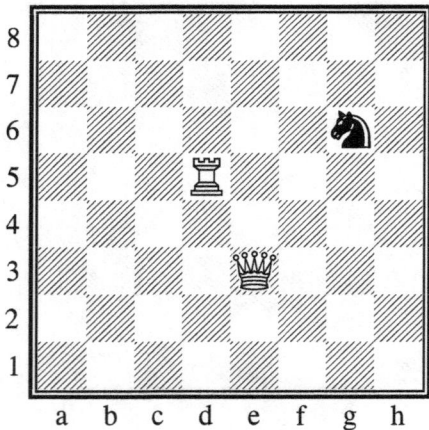

White has _____ product points.

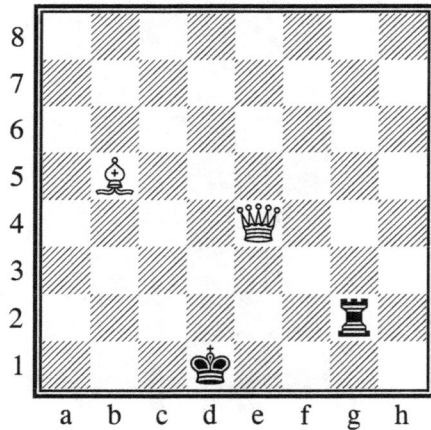

Product using image process

White has _____ product points.

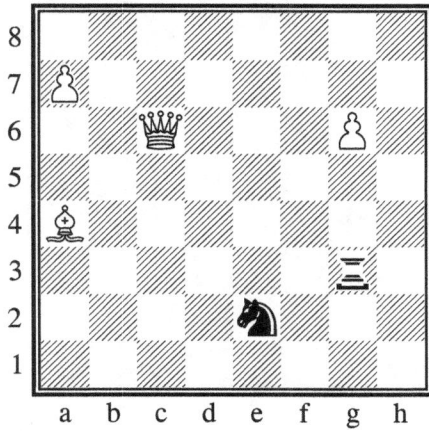

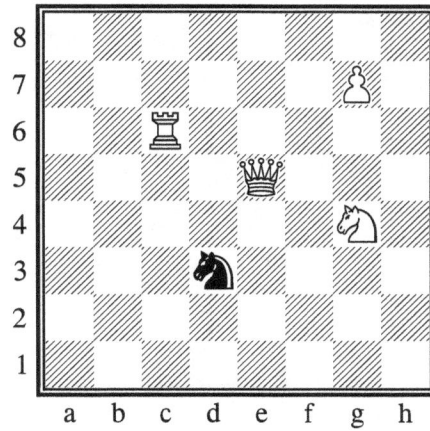

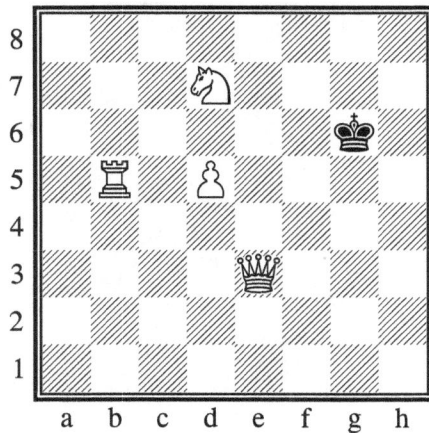

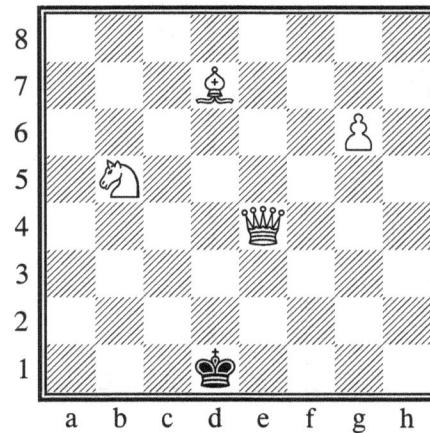

White has _____ product points.

White has _____ product points.

White has _____ product points.

2007 - 2018 © Frank Ho, Amanda Ho, All rights reserved.　　www.homathchess.com

Product using image process

White has _____ product points.

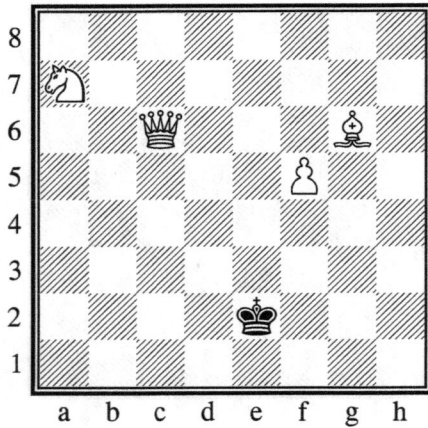

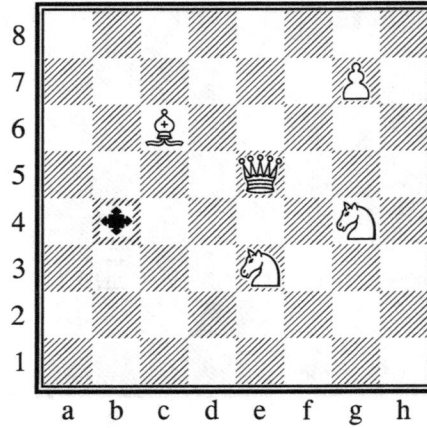

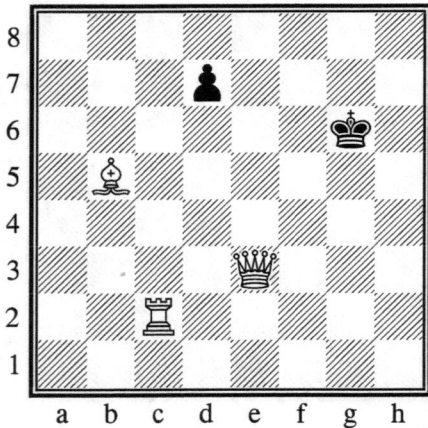

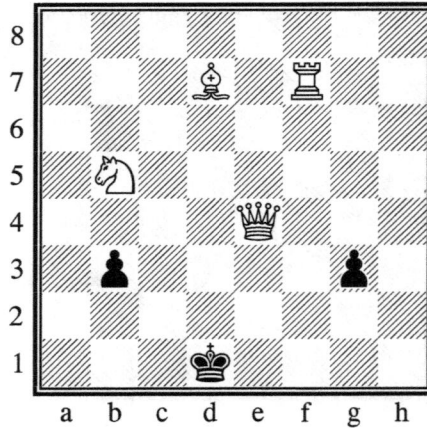

White has _____ product points.

White has _____ product points.

White has _____ product points.

Mom! I Learn Multiplication Using Math-Chess-Puzzles Connection!

Student's Name _____ Date _____

2007 - 2018 © Frank Ho, Amanda Ho, All rights reserved.   www.homathchess.com

Product using image process

White has _____ product points.

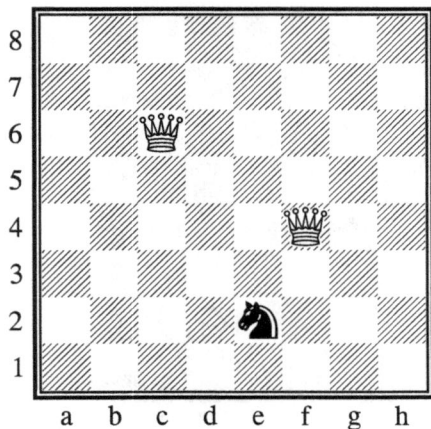

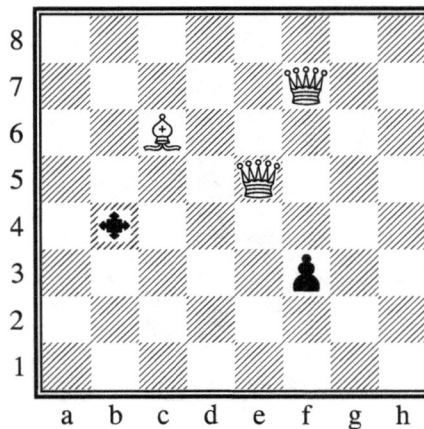

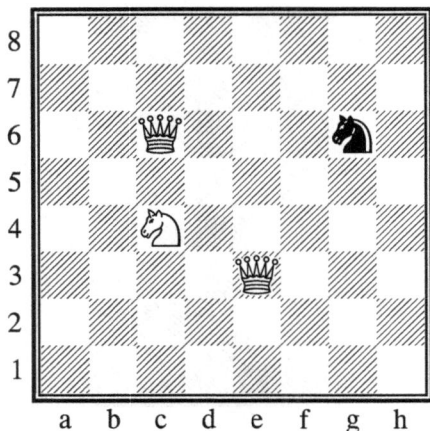

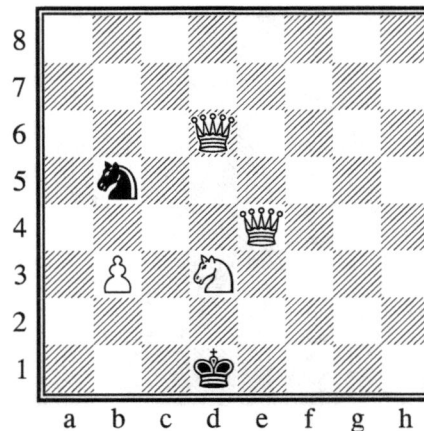

White has _____ product points.

White has _____ product points.

White has _____ product points.

2007 - 2018 © Frank Ho, Amanda Ho, All rights reserved.   www.homathchess.com

Product using image process

White has _____ product points.

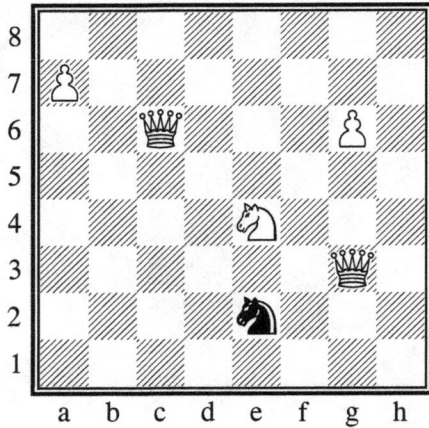

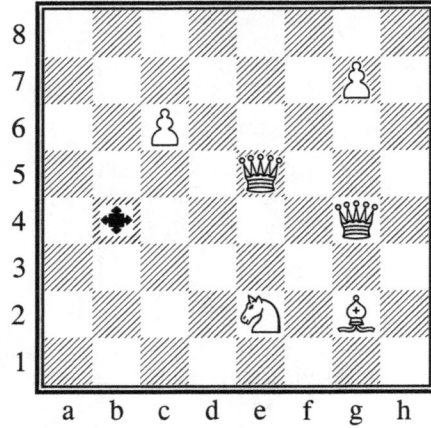

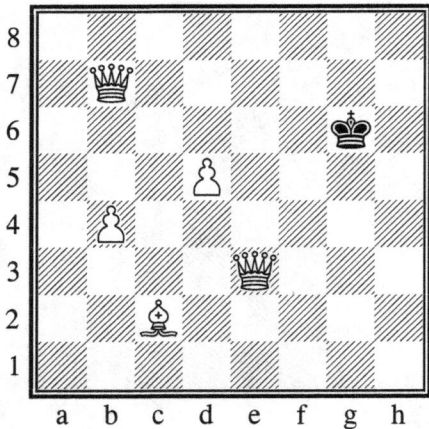

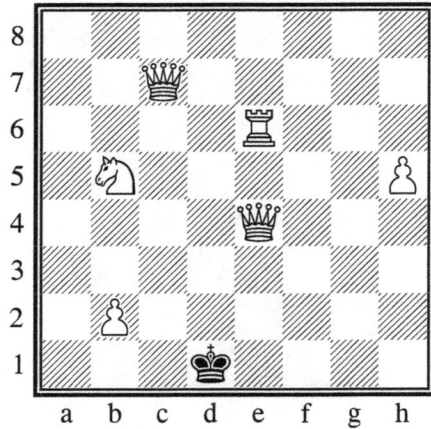

White has _____ product points.

White has _____ product points.

White has _____ product points.

2007 - 2018 © Frank Ho, Amanda Ho, All rights reserved.   www.homathchess.com

Product using image process

White has _____ product points.

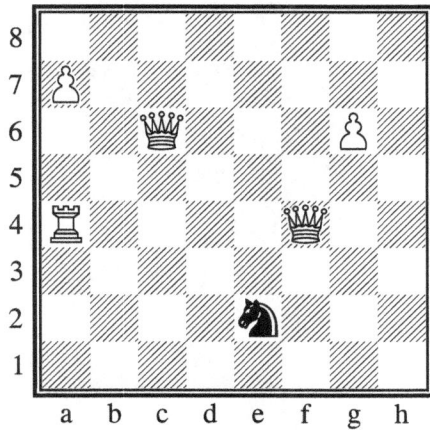

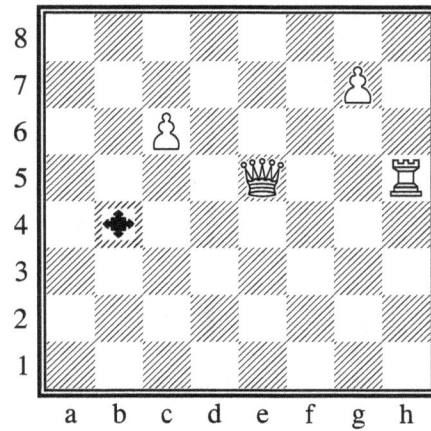

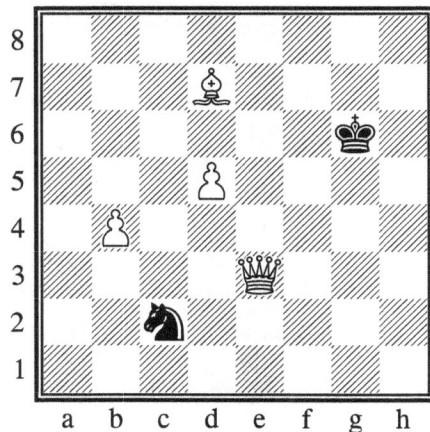

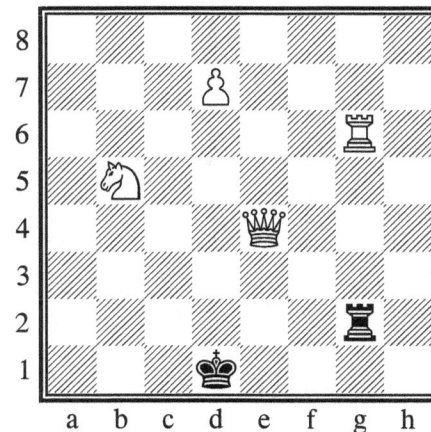

White has _____ product points.

White has _____ product points.

White has _____ product points.

2007 - 2018 © Frank Ho, Amanda Ho, All rights reserved.　　www.homathchess.com

Product using image process

White has _____ product points.

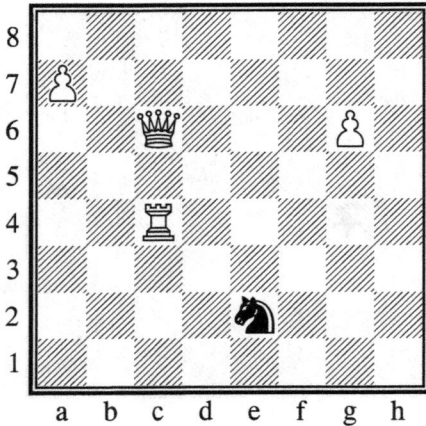

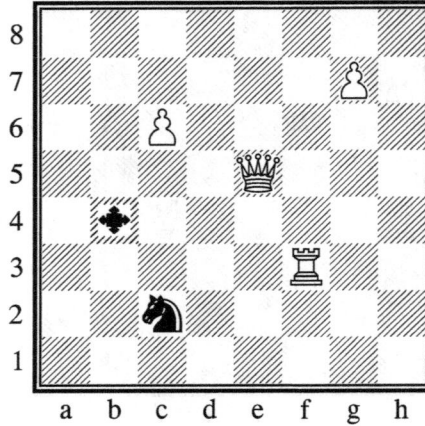

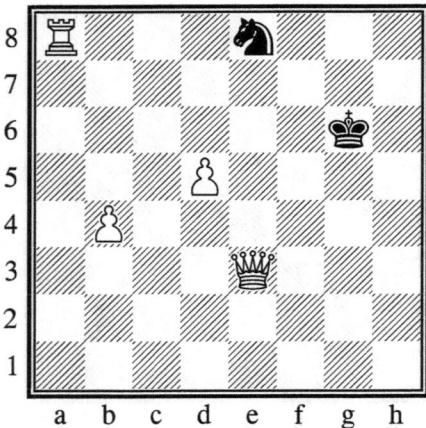

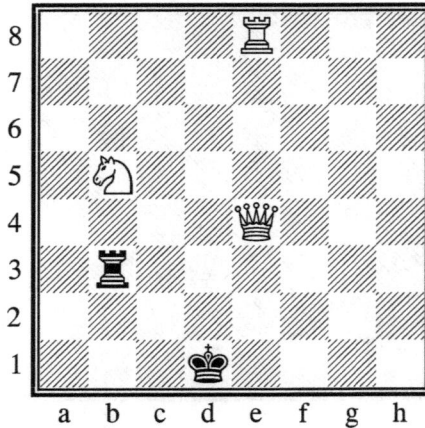

White has _____ product points.

White has _____ product points.

White has _____ product points.

2007 - 2018 © Frank Ho, Amanda Ho, All rights reserved.   www.homathchess.com

Multiplying by relating

Ho Math Chess 何数棋谜 妈!我会棋谜式乘法啦!

Mom! I Learn Multiplication Using Math-Chess-Puzzles Connection!

Student's Name _____ Date _____

2007 - 2018 © Frank Ho, Amanda Ho, All rights reserved.   www.homathchess.com

Multiplying by relating

2007 - 2018 © Frank Ho, Amanda Ho, All rights reserved.   www.homathchess.com

Multiplying by relating

2007 - 2018 © Frank Ho, Amanda Ho, All rights reserved.   www.homathchess.com

Multiplying by relating

Ho Math Chess 何数棋谜 妈!我会棋谜式乘法啦!

Mom! I Learn Multiplication Using Math-Chess-Puzzles Connection!

Student's Name _____ Date _____

2007 - 2018 © Frank Ho, Amanda Ho, All rights reserved.   www.homathchess.com

Multiplying by relating

♙ × 2 = □ × ♙ = □	2 × ♝ = □ 2 × ♙ = □	♝ × 4 = □ ♝ × ♙ = □
4 × ♜ = □ 4 × 1 = □	♜ × 6 = □ ♜ × 1 = □	6 × 7 = □ 6 × ♙ = □
7 × 8 = □ 7 × ♙ = □	8 × ♛ = □ 8 × 1 = □	8 × ♛ = □ 8 × ♙ = □

Multiplying by relating

↓× ♔ ×↓ ♙ × **2** □ ♙ □ × ♙ □	↓× ♔ ×↓ **2** × ♝ □ **2** □ × ♙ □	↓× ♔ ×↓ ♝ × **4** □ ♝ □ × ♙ □
↓× ♔ ×↓ **4** × ♖ □ **4** □ × **1** □	↓× ♔ ×↓ ♖ × **6** □ ♖ □ × **1** □	↓× ♔ ×↓ **6** × **7** □ **6** □ × ♙ □
↓× ♔ ×↓ **7** × **8** □ **7** □ × ♙ □	↓× ♔ ×↓ **8** × ♛ □ **8** □ × **1** □	↓× ♔ ×↓ **8** × ♛ □ **8** □ × ♙ □

2007 - 2018 © Frank Ho, Amanda Ho, All rights reserved.   www.homathchess.com

Multiplying by relating

2007 - 2018 © Frank Ho, Amanda Ho, All rights reserved.   www.homathchess.com

Multiplying by relating

Puzzle 1	Puzzle 2	Puzzle 3
⌐× ♕ ×⌐ ♖ × 2 □ ♖ □ × ♙ □	⌐× ♕ ×⌐ ♖ × ♖ □ 2 □ × ♙ □	⌐× ♕ ×⌐ ♖ × ♖ □ ♖ □ × ♙ □
⌐× 9 ×⌐ ♖ × ♖ □ 4 □ × 1 □	⌐× ♕ ×⌐ ♖ × 6 □ ♖ □ × 1 □	⌐× 9 ×⌐ ♖ × 7 □ 6 □ × ♙ □
⌐× ♕ ×⌐ ♖ × ♖ □ 7 □ × ♙ □	⌐× 9 ×⌐ ♖ × ♖ □ 8 □ × 1 □	⌐× 8 ×⌐ ♖ × ♖ □ 8 □ × ♙ □

Student's Name _____   Date _____

2007 - 2018 © Frank Ho, Amanda Ho, All rights reserved.   www.homathchess.com

Multiplying by relating

⌐× 🨣 ×⌐ 🨣 × 2 □ 🨣 × ♙ □	⌐× 🨣 ×⌐ ♘ × 🨝 □ 2 □ × ♙ □	⌐× 🨣 ×⌐ 🨝 × 4 □ □ 🨝 × ♙ □
⌐× 9 ×⌐ 4 × 🨜 □ □ 🨣 × 1 □	⌐× ♕ ×⌐ 🨜 × 6 □ □ 🨣 × 1 □	⌐× 9 ×⌐ 6 × 7 □ □ 🨣 × ♙ □
⌐× ♕ ×⌐ 7 × 8 □ □ 🨣 × ♙ □	⌐× 9 ×⌐ 8 × ♕ □ □ 🨣 × 1 □	⌐× 8 ×⌐ 8 × ♕ □ □ 🨣 × ♙ □

Ho Math Chess   何数棋谜 妈！我会棋谜式乘法啦！
Mom! I Learn Multiplication Using Math-Chess-Puzzles Connection!

Student's Name _____ Date _____

2007 - 2018 © Frank Ho, Amanda Ho, All rights reserved.   www.homathchess.com

What is the product of all attacking pieces on the ▨ or ☒?

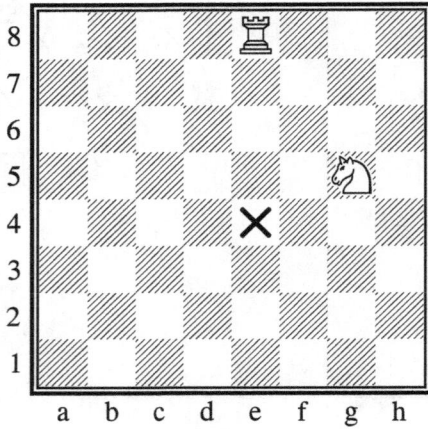

Answer _____

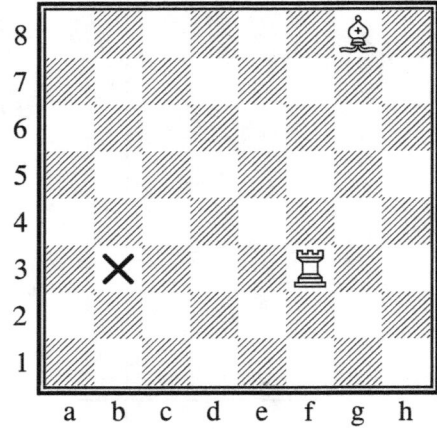

Answer _____

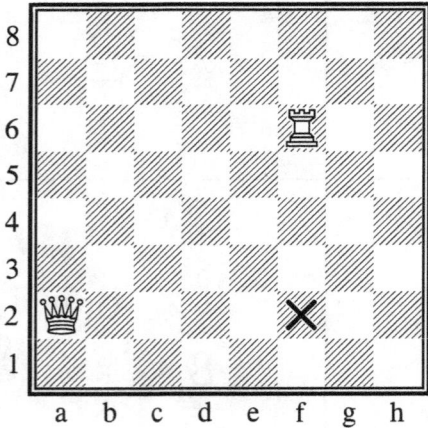

Answer _____

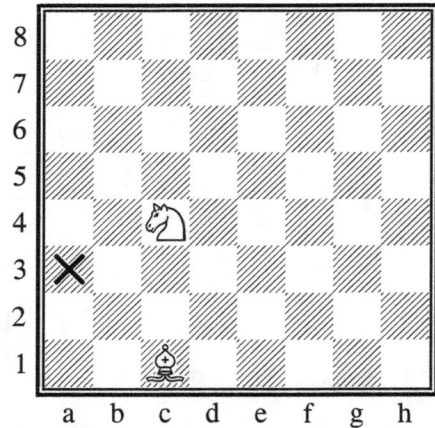

Answer _____

2007 - 2018 © Frank Ho, Amanda Ho, All rights reserved.   www.homathchess.com

What is the product of all attacking pieces on the ✖ or ☒?

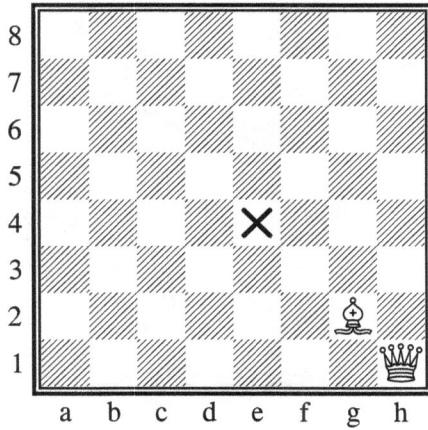

Answer _____

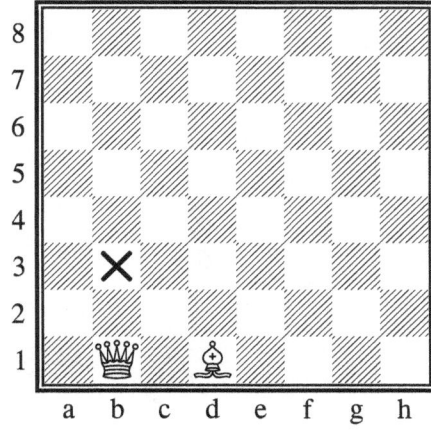

Answer _____

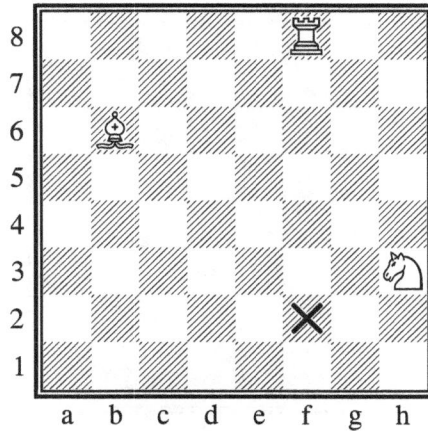

Answer _____

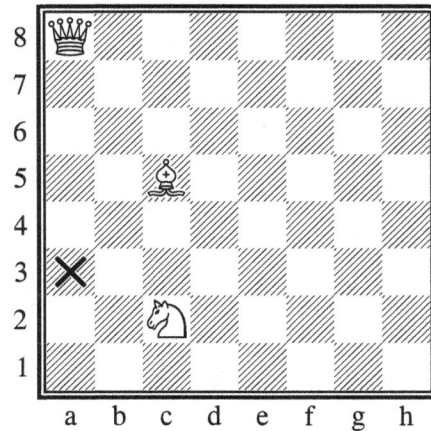

Answer _____

Ho Math Chess 何数棋谜   妈！我会棋谜式乘法啦！
Mom! I Learn Multiplication Using Math-Chess-Puzzles Connection!

Student's Name _____ Date _____

2007 - 2018 © Frank Ho, Amanda Ho, All rights reserved.   www.homathchess.com

What is the product of all attacking pieces on the ✖ or ☒?

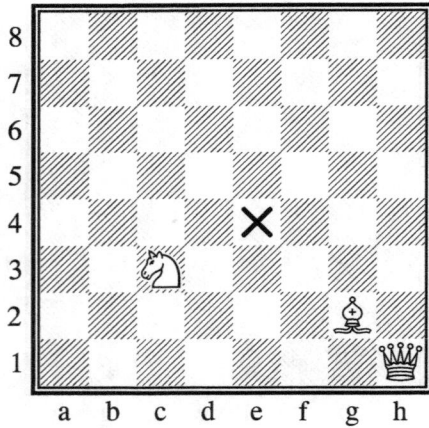

Answer _____

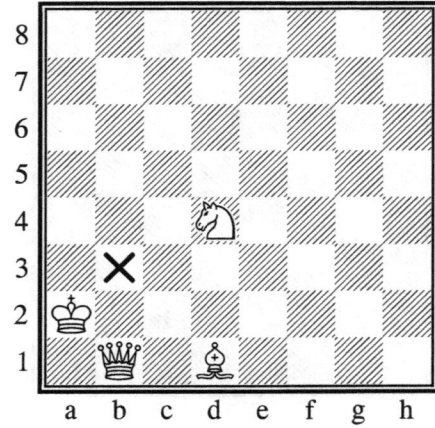

Answer _____

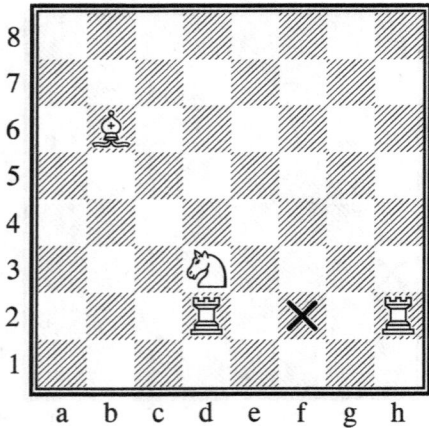

Answer _____

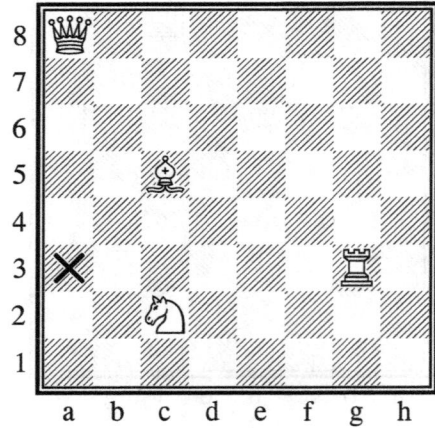

Answer _____

2007 - 2018 © Frank Ho, Amanda Ho, All rights reserved.   www.homathchess.com

What is the product of all attacking pieces on the ✕ or ✕?

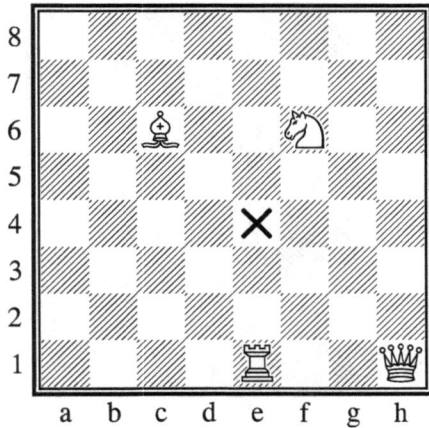

Answer _____

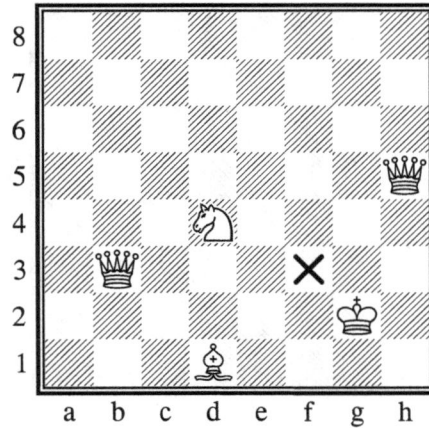

Answer _____

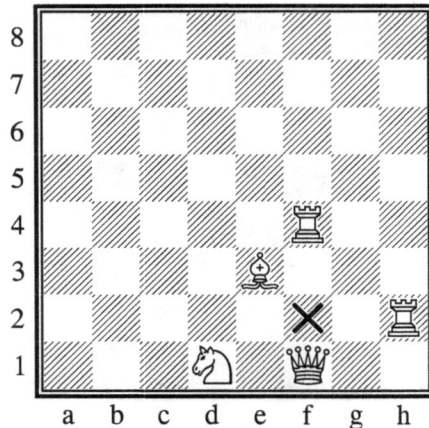

Answer _____

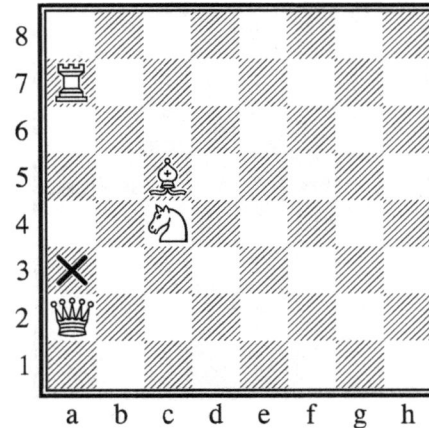

Answer _____

2007 - 2018 © Frank Ho, Amanda Ho, All rights reserved.   www.homathchess.com

What is the product of all attacking pieces on the ✕ or ✕?

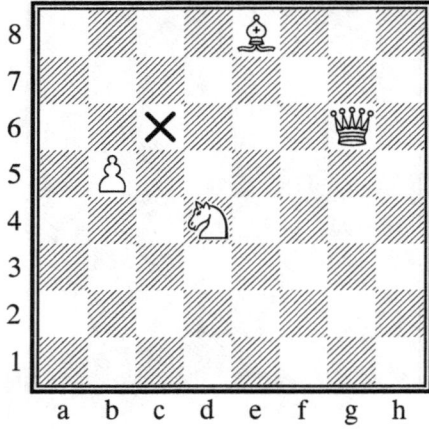

Answer _____

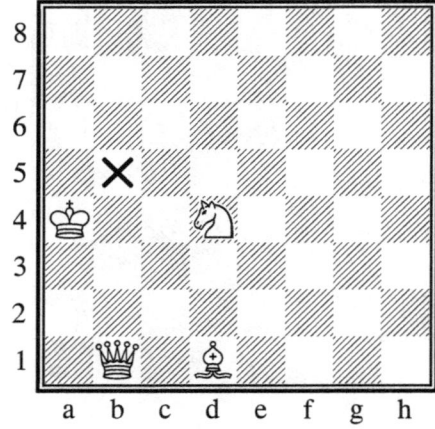

Answer _____

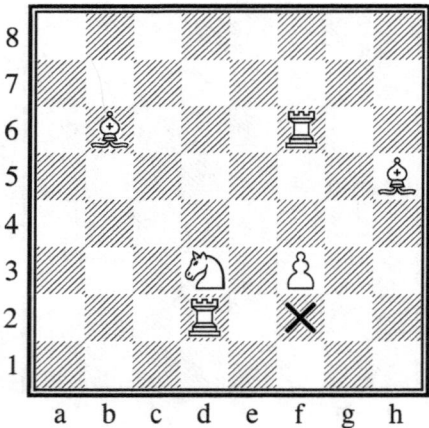

Answer _____

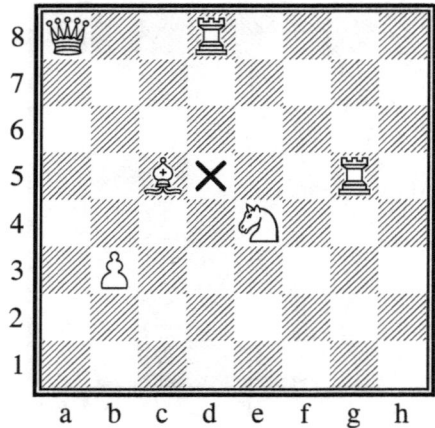

Answer _____

2007 - 2018 © Frank Ho, Amanda Ho, All rights reserved.   www.homathchess.com

What is the product of all attacking pieces on the ✖ or ☒?

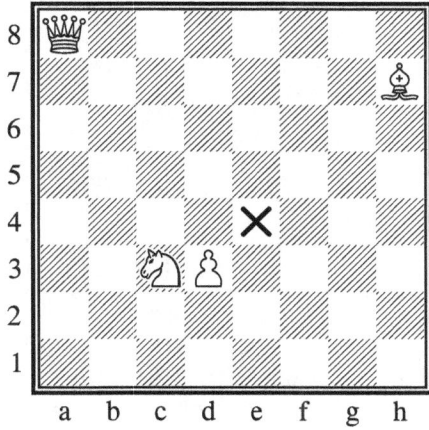

Answer _____

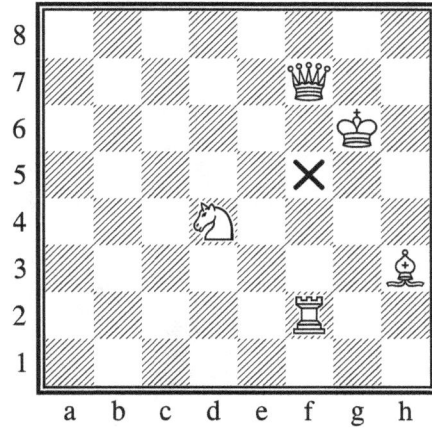

Answer _____

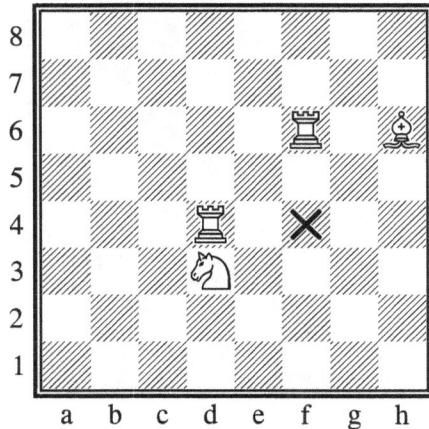

Answer _____

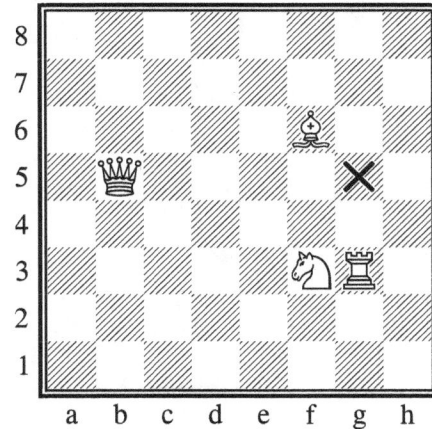

Answer _____

2007 - 2018 © Frank Ho, Amanda Ho, All rights reserved.   www.homathchess.com

What is the product of all attacking pieces on the ✕ or ☒?

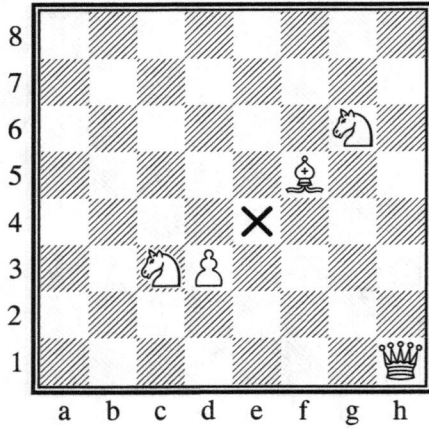

Answer _____

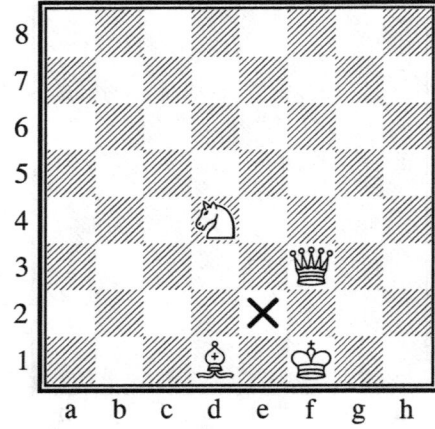

Answer _____

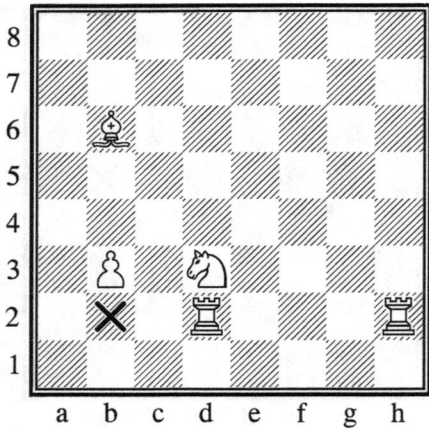

Answer _____

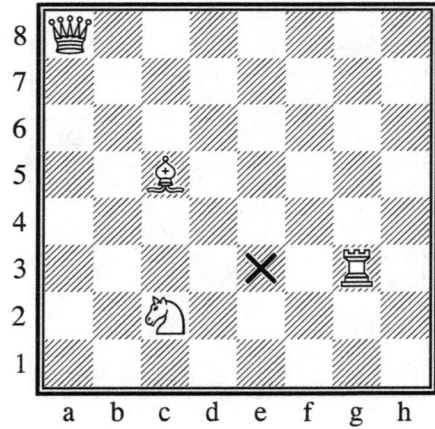

Answer _____

2007 - 2018 © Frank Ho, Amanda Ho, All rights reserved.   www.homathchess.com

What is the product of all attacking pieces on the ✖ or ☒?

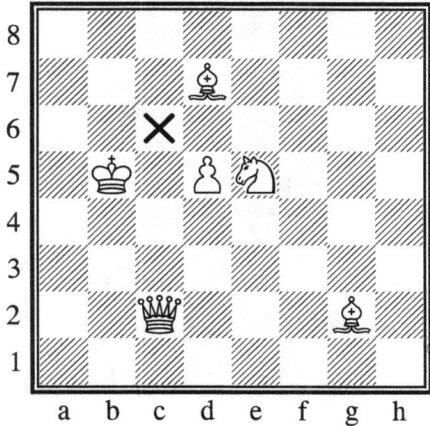

Answer _____

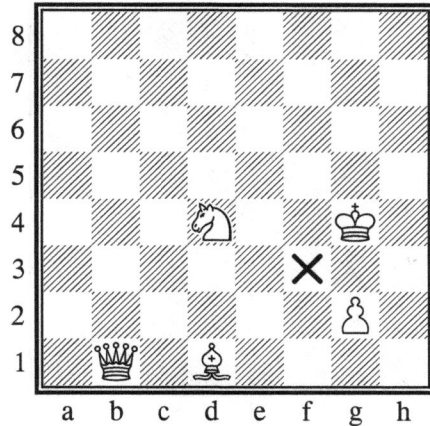

Answer _____

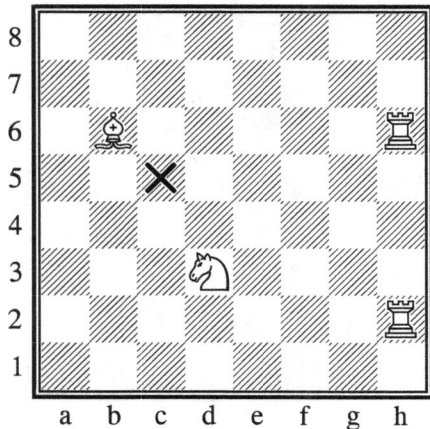

Answer _____

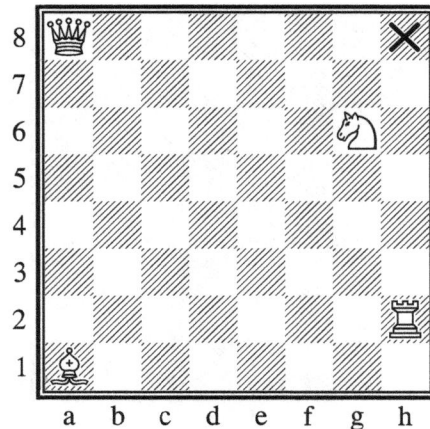

Answer _____

2007 - 2018 © Frank Ho, Amanda Ho, All rights reserved.   www.homathchess.com

What is the product of all attacking pieces on the ✖ or ☒?

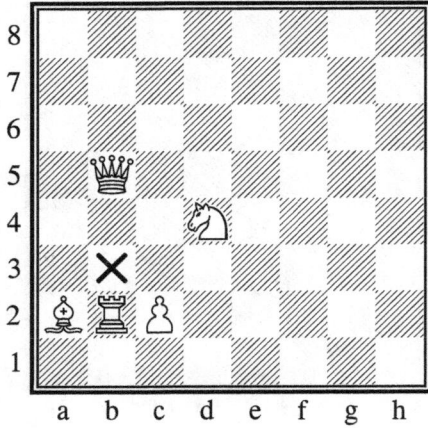

Answer _____

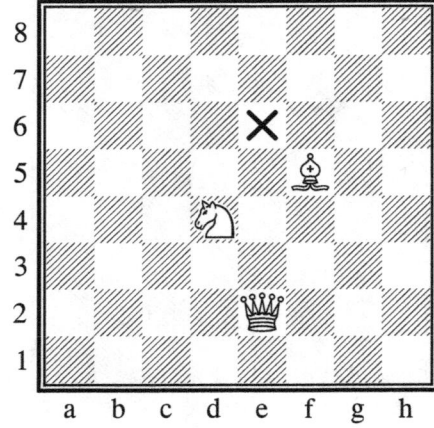

Answer _____

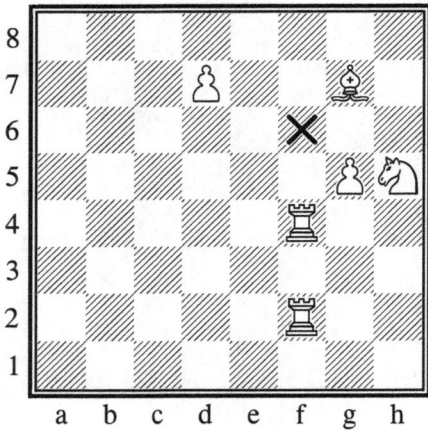

Answer _____

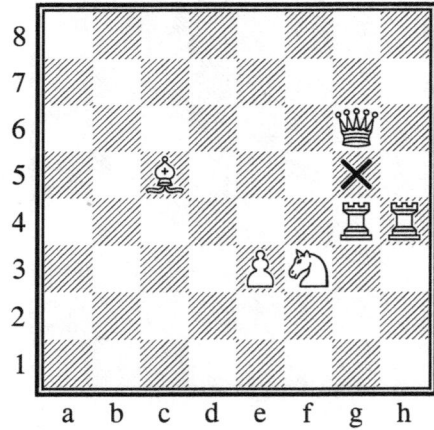

Answer _____

2007 - 2018 © Frank Ho, Amanda Ho, All rights reserved.   www.homathchess.com

What is the product of all attacking pieces on the ✗ or ☒?

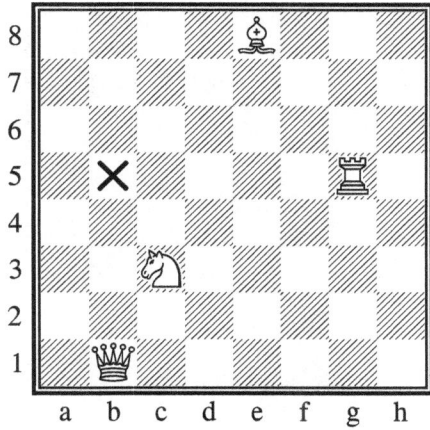

Answer _____

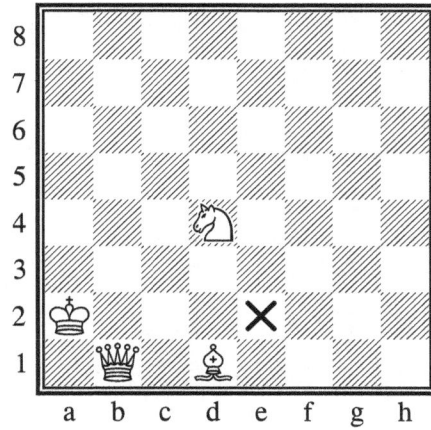

Answer _____

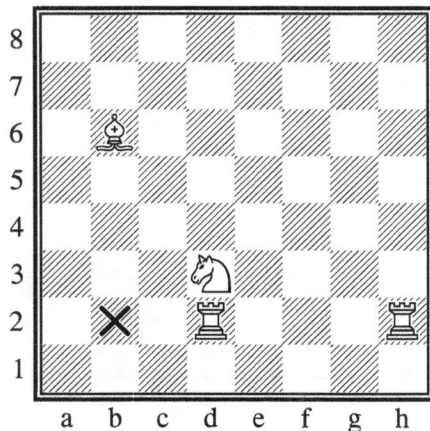

Answer _____

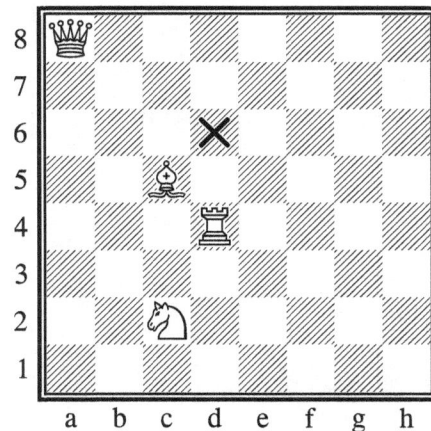

Answer _____

Ho Math Chess   何数棋谜   妈！我会棋谜式乘法啦！
Mom! I Learn Multiplication Using Math-Chess-Puzzles Connection!

Student's Name _____ Date _____

2007 - 2018 © Frank Ho, Amanda Ho, All rights reserved.   www.homathchess.com

Mark (✗) on only the common squares controlled by all chess pieces

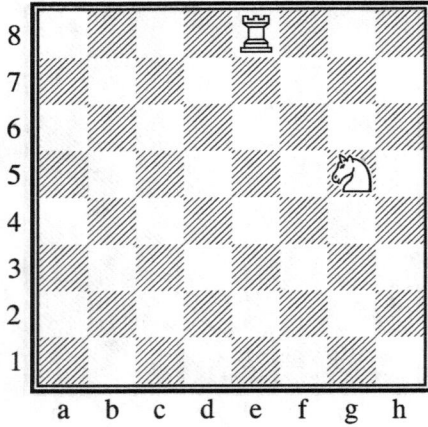

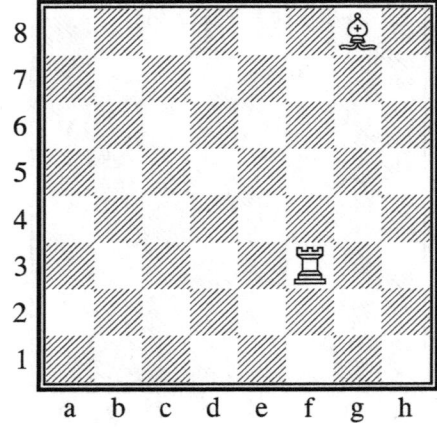

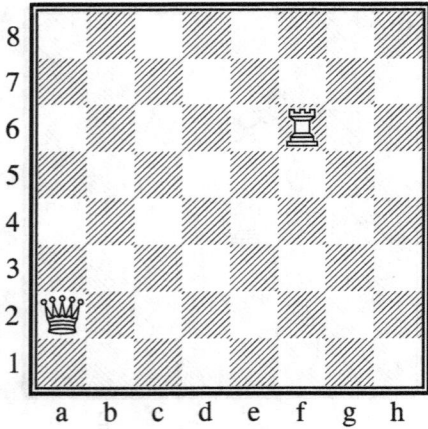

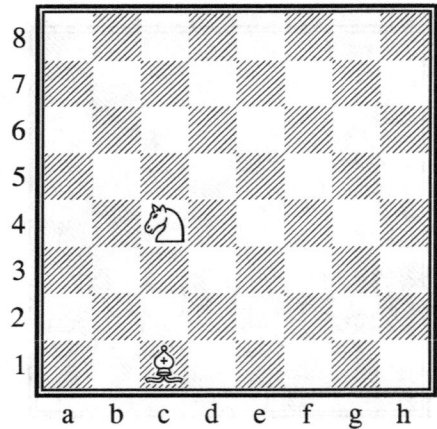

Ho Math Chess 何数棋谜 妈！我会棋谜式乘法啦！
Mom! I Learn Multiplication Using Math-Chess-Puzzles Connection!

Student's Name _____ Date _____

2007 - 2018 © Frank Ho, Amanda Ho, All rights reserved.   www.homathchess.com

Mark (✗) on only the common squares controlled by all chess pieces

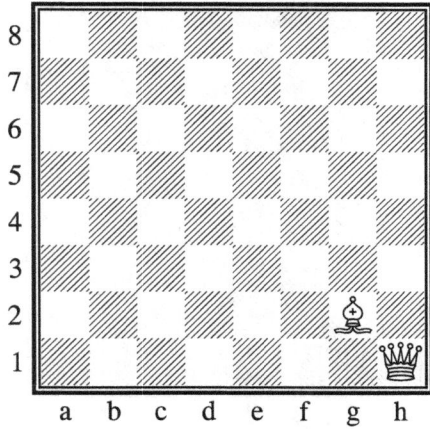

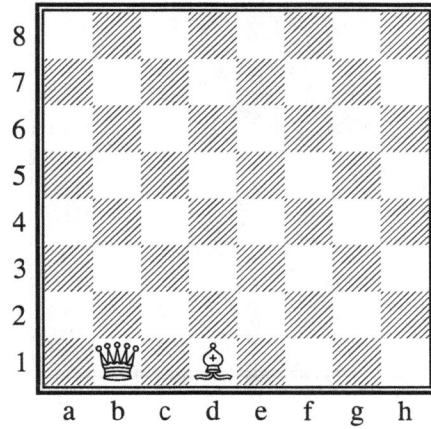

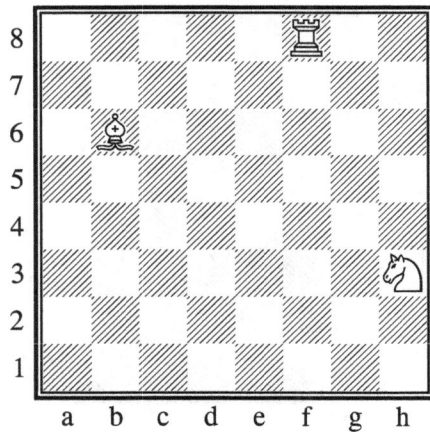

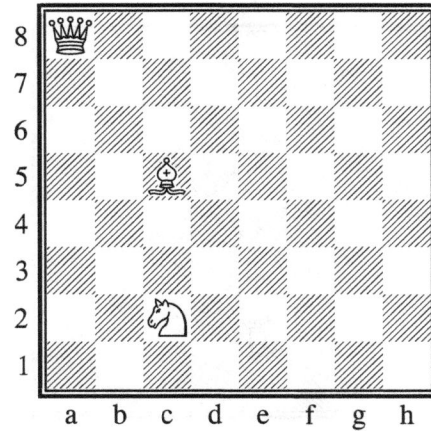

Student's Name _____ Date _____

2007 - 2018 © Frank Ho, Amanda Ho, All rights reserved.   www.homathchess.com

Mark (✗) on only the common squares controlled by all chess pieces

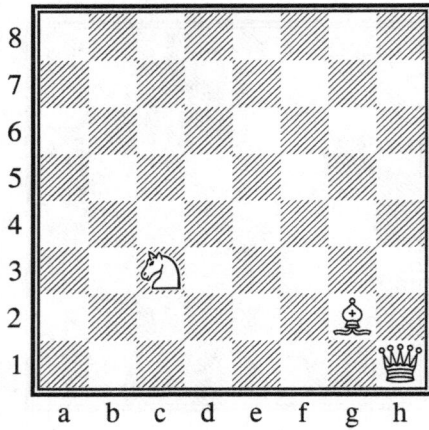

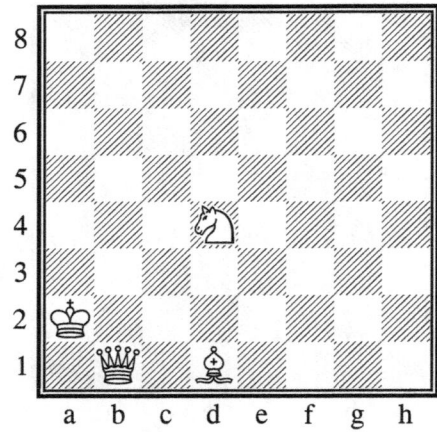

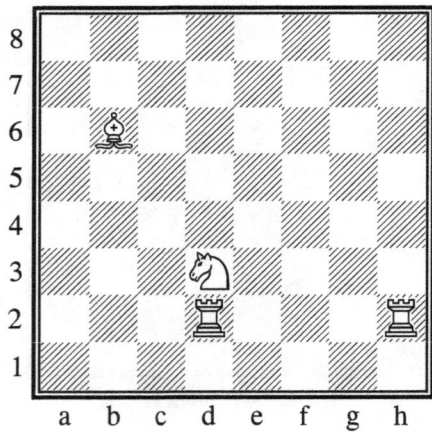

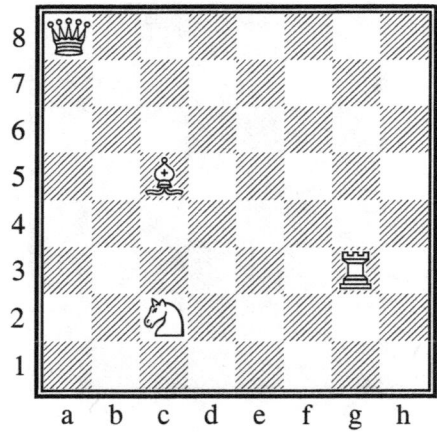

Ho Math Chess 何数棋谜 妈！我会棋谜式乘法啦！
Mom! I Learn Multiplication Using Math-Chess-Puzzles Connection!

Student's Name _____ Date _____

2007 - 2018 © Frank Ho, Amanda Ho, All rights reserved.   www.homathchess.com

Multiplication table

×	0	1	2	3	4	5	6	7	8	9
1										
2										
3										
4										
5										
6										
7										
8										
9										

2007 - 2018 © Frank Ho, Amanda Ho, All rights reserved.   www.homathchess.com

Multiplication table

×	0	1	3	5	7	9	2	4	6	8
2										
4										
6										
8										
1										
3										
5										
7										
9										

2007 - 2018 © Frank Ho, Amanda Ho, All rights reserved.   www.homathchess.com

Multiplication table

Place a "/" over every multiple of 2 and "\" over every multiple of 3.

1	2	3	4	5	6	7	8	9	10
11	12	13	14	15	16	17	18	19	20
21	22	23	24	25	26	27	28	29	30
31	32	33	34	35	36	37	38	39	40
41	42	43	44	45	46	47	48	49	50
51	52	53	54	55	56	57	58	59	60
61	62	63	64	65	66	67	68	69	70
71	72	73	74	75	76	77	78	79	80
81	82	83	84	85	86	87	88	89	90
91	92	93	94	95	96	97	98	99	100

What are the unit digits of 2's multiples ? _____.

The common multiples of 2 and 3 are listed as follows:

_____

The common multiples of 2 and 3 are divisible by _____.

2007 - 2018 © Frank Ho, Amanda Ho, All rights reserved.   www.homathchess.com

Multiplication table

Place a "/" over every multiple of 2 and "\\" over every multiple of 4.

1	2	3	4	5	6	7	8	9	10
11	12	13	14	15	16	17	18	19	20
21	22	23	24	25	26	27	28	29	30
31	32	33	34	35	36	37	38	39	40
41	42	43	44	45	46	47	48	49	50
51	52	53	54	55	56	57	58	59	60
61	62	63	64	65	66	67	68	69	70
71	72	73	74	75	76	77	78	79	80
81	82	83	84	85	86	87	88	89	90
91	92	93	94	95	96	97	98	99	100

The common multiples of 2 and 4 are listed as follows:

_____

Are the above common multiples also the 4's multiples? _____

Mom! I Learn Multiplication Using Math-Chess-Puzzles Connection!

Student's Name _____ Date _____

2007 - 2018 © Frank Ho, Amanda Ho, All rights reserved.　　www.homathchess.com

Pattern (Preparation for equivalent fraction)

$\dfrac{1}{2}=\dfrac{2}{\square}$	$\dfrac{1}{3}=\dfrac{2}{\square}$	$\dfrac{1}{4}=\dfrac{2}{\square}$	$\dfrac{1}{5}=\dfrac{2}{\square}$	$\dfrac{1}{6}=\dfrac{2}{\square}$
$\dfrac{1}{2}=\dfrac{3}{\square}$	$\dfrac{1}{3}=\dfrac{3}{\square}$	$\dfrac{1}{4}=\dfrac{3}{\square}$	$\dfrac{1}{5}=\dfrac{3}{\square}$	$\dfrac{1}{6}=\dfrac{3}{\square}$
$\dfrac{1}{2}=\dfrac{4}{\square}$	$\dfrac{1}{3}=\dfrac{4}{\square}$	$\dfrac{1}{4}=\dfrac{4}{\square}$	$\dfrac{1}{5}=\dfrac{4}{\square}$	$\dfrac{1}{6}=\dfrac{4}{\square}$
$\dfrac{1}{2}=\dfrac{5}{\square}$	$\dfrac{1}{3}=\dfrac{5}{\square}$	$\dfrac{1}{4}=\dfrac{5}{\square}$	$\dfrac{1}{5}=\dfrac{5}{\square}$	$\dfrac{1}{6}=\dfrac{5}{\square}$
$\dfrac{1}{2}=\dfrac{6}{\square}$	$\dfrac{1}{3}=\dfrac{6}{\square}$	$\dfrac{1}{4}=\dfrac{6}{\square}$	$\dfrac{1}{5}=\dfrac{6}{\square}$	$\dfrac{1}{6}=\dfrac{6}{\square}$
$\dfrac{1}{2}=\dfrac{7}{\square}$	$\dfrac{1}{3}=\dfrac{7}{\square}$	$\dfrac{1}{4}=\dfrac{7}{\square}$	$\dfrac{1}{5}=\dfrac{7}{\square}$	$\dfrac{1}{6}=\dfrac{7}{\square}$
$\dfrac{1}{2}=\dfrac{8}{\square}$	$\dfrac{1}{3}=\dfrac{8}{\square}$	$\dfrac{1}{4}=\dfrac{8}{\square}$	$\dfrac{1}{5}=\dfrac{8}{\square}$	$\dfrac{1}{6}=\dfrac{8}{\square}$
$\dfrac{1}{2}=\dfrac{9}{\square}$	$\dfrac{1}{3}=\dfrac{9}{\square}$	$\dfrac{1}{4}=\dfrac{9}{\square}$	$\dfrac{1}{5}=\dfrac{9}{\square}$	$\dfrac{1}{6}=\dfrac{9}{\square}$

Student's Name _____ Date _____

2007 - 2018 © Frank Ho, Amanda Ho, All rights reserved.   www.homathchess.com

Pattern (Preparation for equivalent fraction)

$\dfrac{1}{7}=\dfrac{2}{\square}$	$\dfrac{1}{8}=\dfrac{2}{\square}$	$\dfrac{1}{9}=\dfrac{2}{\square}$	$\dfrac{1}{5}=\dfrac{2}{\square}$	$\dfrac{1}{6}=\dfrac{2}{\square}$
$\dfrac{1}{7}=\dfrac{3}{\square}$	$\dfrac{1}{8}=\dfrac{3}{\square}$	$\dfrac{1}{9}=\dfrac{3}{\square}$	$\dfrac{1}{5}=\dfrac{3}{\square}$	$\dfrac{1}{6}=\dfrac{3}{\square}$
$\dfrac{1}{7}=\dfrac{4}{\square}$	$\dfrac{1}{8}=\dfrac{4}{\square}$	$\dfrac{1}{9}=\dfrac{4}{\square}$	$\dfrac{1}{5}=\dfrac{4}{\square}$	$\dfrac{1}{6}=\dfrac{4}{\square}$
$\dfrac{1}{7}=\dfrac{5}{\square}$	$\dfrac{1}{8}=\dfrac{5}{\square}$	$\dfrac{1}{9}=\dfrac{5}{\square}$	$\dfrac{1}{5}=\dfrac{5}{\square}$	$\dfrac{1}{6}=\dfrac{5}{\square}$
$\dfrac{1}{7}=\dfrac{6}{\square}$	$\dfrac{1}{8}=\dfrac{6}{\square}$	$\dfrac{1}{9}=\dfrac{6}{\square}$	$\dfrac{1}{5}=\dfrac{6}{\square}$	$\dfrac{1}{6}=\dfrac{6}{\square}$
$\dfrac{1}{7}=\dfrac{7}{\square}$	$\dfrac{1}{8}=\dfrac{7}{\square}$	$\dfrac{1}{9}=\dfrac{7}{\square}$	$\dfrac{1}{5}=\dfrac{7}{\square}$	$\dfrac{1}{6}=\dfrac{7}{\square}$
$\dfrac{1}{7}=\dfrac{8}{\square}$	$\dfrac{1}{8}=\dfrac{8}{\square}$	$\dfrac{1}{9}=\dfrac{8}{\square}$	$\dfrac{1}{5}=\dfrac{8}{\square}$	$\dfrac{1}{6}=\dfrac{8}{\square}$
$\dfrac{1}{7}=\dfrac{9}{\square}$	$\dfrac{1}{8}=\dfrac{9}{\square}$	$\dfrac{1}{9}=\dfrac{9}{\square}$	$\dfrac{1}{5}=\dfrac{9}{\square}$	$\dfrac{1}{6}=\dfrac{9}{\square}$

Ho Math Chess 何数棋谜 妈!我会棋谜式乘法啦!

Mom! I Learn Multiplication Using Math-Chess-Puzzles Connection!

Student's Name _____ Date _____

2007 - 2018 © Frank Ho, Amanda Ho, All rights reserved.   www.homathchess.com

Pattern (Preparation for equivalent fraction)

$\frac{1}{2} = \frac{2}{\square}$	$\frac{1}{3} = \frac{2}{\square}$	$\frac{1}{4} = \frac{2}{\square}$	$\frac{1}{5} = \frac{2}{\square}$	$\frac{1}{6} = \frac{2}{\square}$
$\frac{1}{2} = \frac{3}{\square}$	$\frac{1}{3} = \frac{3}{\square}$	$\frac{1}{4} = \frac{3}{\square}$	$\frac{1}{5} = \frac{3}{\square}$	$\frac{1}{6} = \frac{3}{\square}$
$\frac{1}{2} = \frac{4}{\square}$	$\frac{1}{3} = \frac{4}{\square}$	$\frac{1}{4} = \frac{4}{\square}$	$\frac{1}{5} = \frac{4}{\square}$	$\frac{1}{6} = \frac{4}{\square}$
$\frac{1}{2} = \frac{5}{\square}$	$\frac{1}{3} = \frac{5}{\square}$	$\frac{1}{4} = \frac{5}{\square}$	$\frac{1}{5} = \frac{5}{\square}$	$\frac{1}{6} = \frac{5}{\square}$
$\frac{1}{2} = \frac{6}{\square}$	$\frac{1}{3} = \frac{6}{\square}$	$\frac{1}{4} = \frac{6}{\square}$	$\frac{1}{5} = \frac{6}{\square}$	$\frac{1}{6} = \frac{6}{\square}$
$\frac{1}{2} = \frac{7}{\square}$	$\frac{1}{3} = \frac{7}{\square}$	$\frac{1}{4} = \frac{7}{\square}$	$\frac{1}{5} = \frac{7}{\square}$	$\frac{1}{6} = \frac{7}{\square}$
$\frac{1}{2} = \frac{8}{\square}$	$\frac{1}{3} = \frac{8}{\square}$	$\frac{1}{4} = \frac{8}{\square}$	$\frac{1}{5} = \frac{8}{\square}$	$\frac{1}{6} = \frac{8}{\square}$

Mom! I Learn Multiplication Using Math-Chess-Puzzles Connection!

Student's Name _____ Date _____

2007 - 2018 © Frank Ho, Amanda Ho, All rights reserved.   www.homathchess.com

Learning multiplication with multi-concept and multi-format

c	9	8	7
b	2	3	6
a	3	4	5
	1	2	3

The original square is at b2 = □.

Student's Name _____   Date _____

2007 - 2018 © Frank Ho, Amanda Ho, All rights reserved.   www.homathchess.com

Learning multiplication with multi-concept and multi-format

c	9	8	7
b	2	3	6
a	3	4	5
	1	2	3

The original square is at b2 = □.

$$\square \times \triangle = \bigcirc, \quad \bigcirc = \triangle \times \square, \quad \square \overline{)\dfrac{\triangle}{\bigcirc}}, \quad \triangle \overline{)\dfrac{\square}{\bigcirc}}, \quad \square \dfrac{)\bigcirc}{\triangle},$$

$$\triangle \dfrac{)\bigcirc}{\square}$$

$$\square = \bigcirc \times \dfrac{1}{\triangle}, \quad \triangle = \bigcirc \times \dfrac{1}{\square}, \quad \dfrac{1}{\square} = \dfrac{\triangle}{\bigcirc}, \quad \dfrac{1}{\triangle} = \dfrac{\square}{\bigcirc}, \quad \dfrac{\bigcirc}{\triangle} = \dfrac{\square}{1},$$

$$\dfrac{\bigcirc}{\square} = \dfrac{\triangle}{1}$$

$$\square \times \triangle = \bigcirc, \quad \bigcirc = \triangle \times \square, \quad \square \overline{)\dfrac{\triangle}{\bigcirc}}, \quad \triangle \overline{)\dfrac{\square}{\bigcirc}}, \quad \square \dfrac{)\bigcirc}{\triangle},$$

$$\triangle \dfrac{)\bigcirc}{\square}$$

$$\square = \bigcirc \times \dfrac{1}{\triangle}, \quad \triangle = \bigcirc \times \dfrac{1}{\square}, \quad \dfrac{1}{\square} = \dfrac{\triangle}{\bigcirc}, \quad \dfrac{1}{\triangle} = \dfrac{\square}{\bigcirc}, \quad \dfrac{\bigcirc}{\triangle} = \dfrac{\square}{1},$$

$$\dfrac{\bigcirc}{\square} = \dfrac{\triangle}{1}$$

2007 - 2018 © Frank Ho, Amanda Ho, All rights reserved.   www.homathchess.com

Learning multiplication with multi-concept and multi-format

c	9	8	7
b	2	3	6
a	3	4	5
	1	2	3

The original square is at b2 = ☐.

$$\square \times \triangle = \bigcirc, \quad \bigcirc = \triangle \times \square, \quad \square \overline{)\dfrac{\triangle}{\bigcirc}}, \quad \triangle \overline{)\dfrac{\square}{\bigcirc}}, \quad \dfrac{\square \,)\, \bigcirc}{\triangle},$$

$$\dfrac{\triangle \,)\, \bigcirc}{\square},$$

$$\square = \bigcirc \times \dfrac{1}{\triangle}, \quad \triangle = \bigcirc \times \dfrac{1}{\square}, \quad \dfrac{1}{\square} = \dfrac{\triangle}{\bigcirc}, \quad \dfrac{1}{\triangle} = \dfrac{\square}{\bigcirc}, \quad \dfrac{\bigcirc}{\triangle} = \dfrac{\square}{1},$$

$$\dfrac{\bigcirc}{\square} = \dfrac{\triangle}{1}$$

$$\square \times \triangle = \bigcirc, \quad \bigcirc = \triangle \times \square, \quad \square \overline{)\dfrac{\triangle}{\bigcirc}}, \quad \triangle \overline{)\dfrac{\square}{\bigcirc}},$$

$$\dfrac{\square \,)\, \bigcirc}{\triangle}, \quad \dfrac{\triangle \,)\, \bigcirc}{\square},$$

$$\square = \bigcirc \times \dfrac{1}{\triangle}, \quad \triangle = \bigcirc \times \dfrac{1}{\square}, \quad \dfrac{1}{\square} = \dfrac{\triangle}{\bigcirc}, \quad \dfrac{1}{\triangle} = \dfrac{\square}{\bigcirc}, \quad \dfrac{\bigcirc}{\triangle} = \dfrac{\square}{1},$$

$$\dfrac{\bigcirc}{\square} = \dfrac{\triangle}{1}$$

2007 - 2018 © Frank Ho, Amanda Ho, All rights reserved.   www.homathchess.com

Learning multiplication with multi-concept and multi-format

c	9	8	7
b	2	3	6
a	3	4	5
	1	2	3

The original square is at b2 = □.

Student's Name _____ Date _____

2007 - 2018 © Frank Ho, Amanda Ho, All rights reserved.   www.homathchess.com

Learning multiplication with multi-concept and multi-format

c	9	8	7
b	2	3	6
a	3	4	5
	1	2	3

The original square is at b2 = □.

$$\square \times \triangle = \bigcirc, \quad \bigcirc = \triangle \times \square, \quad \square \overline{)\bigcirc}, \quad \triangle \overline{)\bigcirc}, \quad \square \frac{\square \bigcirc}{\triangle},$$

$$\triangle \frac{\bigcirc}{\square}$$

$$\square = \bigcirc \times \frac{1}{\triangle}, \quad \triangle = \bigcirc \times \frac{1}{\square}, \quad \frac{1}{\square} = \frac{\triangle}{\bigcirc}, \quad \frac{1}{\triangle} = \frac{\square}{\bigcirc}, \quad \frac{\bigcirc}{\triangle} = \frac{\square}{1},$$

$$\frac{\bigcirc}{\square} = \frac{\triangle}{1}$$

2007 - 2018 © Frank Ho, Amanda Ho, All rights reserved.   www.homathchess.com

Learning multiplication with multi-concept and multi-format

c	9	8	7
b	2	4	6
a	3	4	5
	1	2	3

The original square is at b2 = □.

Student's Name _____ Date _____

2007 - 2018 © Frank Ho, Amanda Ho, All rights reserved.　www.homathchess.com

Learning multiplication with multi-concept and multi-format

c	9	8	7
b	2	4	6
a	3	4	5
	1	2	3

The original square is at b2 = □.

$\square \times \triangle = \bigcirc$, $\bigcirc = \triangle \times \square$, $\square \overline{)\dfrac{\triangle}{\bigcirc}}$, $\triangle \overline{)\dfrac{\square}{\bigcirc}}$, $\square \dfrac{\bigcirc}{\triangle}$, $\triangle \dfrac{\dfrac{\bigcirc}{\square}}{}$

$\square = \bigcirc \times \dfrac{1}{\triangle}$, $\triangle = \bigcirc \times \dfrac{1}{\square}$, $\dfrac{1}{\square} = \dfrac{\triangle}{\bigcirc}$, $\dfrac{1}{\triangle} = \dfrac{\square}{\bigcirc}$, $\dfrac{\bigcirc}{\triangle} = \dfrac{\square}{1}$, $\dfrac{\bigcirc}{\square} = \dfrac{\triangle}{1}$

$\square \times \triangle = \bigcirc$, $\bigcirc = \triangle \times \square$, $\square \overline{)\dfrac{\triangle}{\bigcirc}}$, $\triangle \overline{)\dfrac{\square}{\bigcirc}}$, $\square \dfrac{\bigcirc}{\triangle}$, $\triangle \dfrac{\dfrac{\bigcirc}{\square}}{}$

$\square = \bigcirc \times \dfrac{1}{\triangle}$, $\triangle = \bigcirc \times \dfrac{1}{\square}$, $\dfrac{1}{\square} = \dfrac{\triangle}{\bigcirc}$, $\dfrac{1}{\triangle} = \dfrac{\square}{\bigcirc}$, $\dfrac{\bigcirc}{\triangle} = \dfrac{\square}{1}$, $\dfrac{\bigcirc}{\square} = \dfrac{\triangle}{1}$

2007 - 2018 © Frank Ho, Amanda Ho, All rights reserved.   www.homathchess.com

Learning multiplication with multi-concept and multi-format

c	9	8	7
b	2	4	6
a	3	4	5
	1	2	3

The original square is at b2 = □.

Student's Name _____ Date _____

2007 - 2018 © Frank Ho, Amanda Ho, All rights reserved.　　www.homathchess.com

Learning multiplication with multi-concept and multi-format

c	9	8	7
b	2	4	6
a	3	4	5
	1	2	3

The original square is at b2 = □.

$$\square \times \triangle = \bigcirc, \quad \bigcirc = \triangle \times \square, \quad \square\overline{)\bigcirc}\,\dfrac{\triangle}{}, \quad \triangle\overline{)\bigcirc}\,\dfrac{\square}{}, \quad \dfrac{\square\,\bigcirc}{\triangle},$$

$$\dfrac{\triangle\,\bigcirc}{\square}$$

$$\square = \bigcirc \times \dfrac{1}{\triangle}, \quad \triangle = \bigcirc \times \dfrac{1}{\square}, \quad \dfrac{1}{\square} = \dfrac{\triangle}{\bigcirc}, \quad \dfrac{1}{\triangle} = \dfrac{\square}{\bigcirc}, \quad \dfrac{\bigcirc}{\triangle} = \dfrac{\square}{1},$$

$$\dfrac{\bigcirc}{\square} = \dfrac{\triangle}{1}$$

$$\square \times \triangle = \bigcirc, \quad \bigcirc = \triangle \times \square, \quad \square\overline{)\bigcirc}\,\dfrac{\triangle}{}, \quad \triangle\overline{)\bigcirc}\,\dfrac{\square}{}, \quad \dfrac{\square\,\bigcirc}{\triangle},$$

$$\dfrac{\triangle\,\bigcirc}{\square}$$

$$\square = \bigcirc \times \dfrac{1}{\triangle}, \quad \triangle = \bigcirc \times \dfrac{1}{\square}, \quad \dfrac{1}{\square} = \dfrac{\triangle}{\bigcirc}, \quad \dfrac{1}{\triangle} = \dfrac{\square}{\bigcirc}, \quad \dfrac{\bigcirc}{\triangle} = \dfrac{\square}{1},$$

$$\dfrac{\bigcirc}{\square} = \dfrac{\triangle}{1}$$

Student's Name _____ Date _____

2007 - 2018 © Frank Ho, Amanda Ho, All rights reserved.　　www.homathchess.com

Learning multiplication with multi-concept and multi-format

c	9	8	7
b	2	4	6
a	3	4	5
	1	2	3

The original square is at b2 = □.

$$\square \times \triangle = \bigcirc, \quad \bigcirc = \triangle \times \square, \quad \square\overline{)\dfrac{\triangle}{\bigcirc}}, \quad \triangle\overline{)\dfrac{\square}{\bigcirc}}, \quad \square\,)\dfrac{\bigcirc}{\triangle},$$

$$\triangle\,)\dfrac{\bigcirc}{\square}$$

$$\square = \bigcirc \times \dfrac{1}{\triangle}, \quad \triangle = \bigcirc \times \dfrac{1}{\square}, \quad \dfrac{1}{\square} = \dfrac{\triangle}{\bigcirc}, \quad \dfrac{1}{\triangle} = \dfrac{\square}{\bigcirc}, \quad \dfrac{\bigcirc}{\triangle} = \dfrac{\square}{1},$$

$$\dfrac{\bigcirc}{\square} = \dfrac{\triangle}{1}$$

2007 - 2018 © Frank Ho, Amanda Ho, All rights reserved.   www.homathchess.com

Learning multiplication with multi-concept and multi-format

c	9	8	7
b	2	5	6
a	3	4	5
	1	2	3

The original square is at b2 = □.

$$\square \times \leftrightarrow\updownarrow = \_ \times \_ = \_ \qquad \square \times \times = \_ \times \_ = \_$$

$$\square \times \leftrightarrow\updownarrow = \_ \times \_ = \_ \qquad \square \times \times = \_ \times \_ = \_$$

$$\square \times \leftrightarrow\updownarrow = \_ \times \_ = \_ \qquad \square \times \times = \_ \times \_ = \_$$

$$\square \times \leftrightarrow\updownarrow = \_ \times \_ = \_ \qquad \square \times \times = \_ \times \_ = \_$$

$\leftrightarrow\updownarrow$

$$\square \times \triangle = \bigcirc, \quad \bigcirc = \triangle \times \square, \quad \square\sqrt{\bigcirc}, \quad \triangle\sqrt{\bigcirc}, \quad \dfrac{\square}{\triangle}\sqrt{\bigcirc}, \quad \dfrac{\bigcirc}{\square}\sqrt{\triangle},$$

$$\square = \bigcirc \times \dfrac{1}{\triangle}, \quad \triangle = \bigcirc \times \dfrac{1}{\square}, \quad \dfrac{1}{\square} = \dfrac{\triangle}{\bigcirc}, \quad \dfrac{1}{\triangle} = \dfrac{\square}{\bigcirc}, \quad \dfrac{\bigcirc}{\triangle} = \dfrac{\square}{1},$$

$$\dfrac{\bigcirc}{\square} = \dfrac{\triangle}{1}$$

Student's Name _____ Date _____

2007 - 2018 © Frank Ho, Amanda Ho, All rights reserved.　www.homathchess.com

Learning multiplication with multi-concept and multi-format

c	9	8	7
b	2	5	6
a	3	4	5
	1	2	3

The original square is at b2 = □.

$$\square \times \triangle = \bigcirc, \quad \bigcirc = \triangle \times \square, \quad \square \overline{)\dfrac{\triangle}{\bigcirc}}, \quad \triangle \overline{)\dfrac{\square}{\bigcirc}}, \quad \dfrac{\square \, \lfloor \bigcirc}{\triangle},$$

$$\triangle \, \lfloor \dfrac{\bigcirc}{\square}$$

$$\square = \bigcirc \times \dfrac{1}{\triangle}, \quad \triangle = \bigcirc \times \dfrac{1}{\square}, \quad \dfrac{1}{\square} = \dfrac{\triangle}{\bigcirc}, \quad \dfrac{1}{\triangle} = \dfrac{\square}{\bigcirc}, \quad \dfrac{\bigcirc}{\triangle} = \dfrac{\square}{1},$$

$$\dfrac{\bigcirc}{\square} = \dfrac{\triangle}{1}$$

$$\square \times \triangle = \bigcirc, \quad \bigcirc = \triangle \times \square, \quad \square \overline{)\dfrac{\triangle}{\bigcirc}}, \quad \triangle \overline{)\dfrac{\square}{\bigcirc}}, \quad \dfrac{\square \, \lfloor \bigcirc}{\triangle},$$

$$\triangle \, \lfloor \dfrac{\bigcirc}{\square}$$

$$\square = \bigcirc \times \dfrac{1}{\triangle}, \quad \triangle = \bigcirc \times \dfrac{1}{\square}, \quad \dfrac{1}{\square} = \dfrac{\triangle}{\bigcirc}, \quad \dfrac{1}{\triangle} = \dfrac{\square}{\bigcirc}, \quad \dfrac{\bigcirc}{\triangle} = \dfrac{\square}{1},$$

$$\dfrac{\bigcirc}{\square} = \dfrac{\triangle}{1}$$

2007 - 2018 © Frank Ho, Amanda Ho, All rights reserved.　　www.homathchess.com

Learning multiplication with multi-concept and multi-format

c	9	8	7
b	2	5	6
a	3	4	5
	1	2	3

The original square is at b2 = □.

$$\square \times \triangle = \bigcirc, \quad \bigcirc = \triangle \times \square, \quad \square\overline{)\bigcirc}, \quad \triangle\overline{)\bigcirc}, \quad \dfrac{\square}{\dfrac{\bigcirc}{\triangle}},$$

$$\dfrac{\triangle}{\dfrac{\bigcirc}{\square}}$$

$$\square = \bigcirc \times \dfrac{1}{\triangle}, \quad \triangle = \bigcirc \times \dfrac{1}{\square}, \quad \dfrac{1}{\square} = \dfrac{\triangle}{\bigcirc}, \quad \dfrac{1}{\triangle} = \dfrac{\square}{\bigcirc}, \quad \dfrac{\bigcirc}{\triangle} = \dfrac{\square}{1},$$

$$\dfrac{\bigcirc}{\square} = \dfrac{\triangle}{1}$$

$$\square \times \triangle = \bigcirc, \quad \bigcirc = \triangle \times \square, \quad \square\overline{)\bigcirc}, \quad \triangle\overline{)\bigcirc}, \quad \dfrac{\square}{\dfrac{\bigcirc}{\triangle}},$$

$$\dfrac{\triangle}{\dfrac{\bigcirc}{\square}}$$

$$\square = \bigcirc \times \dfrac{1}{\triangle}, \quad \triangle = \bigcirc \times \dfrac{1}{\square}, \quad \dfrac{1}{\square} = \dfrac{\triangle}{\bigcirc}, \quad \dfrac{1}{\triangle} = \dfrac{\square}{\bigcirc}, \quad \dfrac{\bigcirc}{\triangle} = \dfrac{\square}{1},$$

$$\dfrac{\bigcirc}{\square} = \dfrac{\triangle}{1}$$

Mom! I Learn Multiplication Using Math-Chess-Puzzles Connection!

Student's Name _____ Date _____

2007 - 2018 © Frank Ho, Amanda Ho, All rights reserved.　　www.homathchess.com

Learning multiplication with multi-concept and multi-format

c	9	8	7
b	2	5	6
a	3	4	5
	1	2	3

The original square is at b2 = □.

2007 - 2018 © Frank Ho, Amanda Ho, All rights reserved.   www.homathchess.com

Learning multiplication with multi-concept and multi-format

c	9	8	7
b	2	5	6
a	3	4	5
	1	2	3

The original square is at b2 = □.

$$\square \times \triangle = \bigcirc, \quad \bigcirc = \triangle \times \square, \quad \square \overline{)\bigcirc}, \quad \triangle \overline{)\bigcirc}, \quad \square \dfrac{)\bigcirc}{\triangle},$$

$$\triangle \dfrac{)\bigcirc}{\square}$$

$$\square = \bigcirc \times \dfrac{1}{\triangle}, \quad \triangle = \bigcirc \times \dfrac{1}{\square}, \quad \dfrac{1}{\square} = \dfrac{\triangle}{\bigcirc}, \quad \dfrac{1}{\triangle} = \dfrac{\square}{\bigcirc}, \quad \dfrac{\bigcirc}{\triangle} = \dfrac{\square}{1},$$

$$\dfrac{\bigcirc}{\square} = \dfrac{\triangle}{1}$$

Ho Math Chess   何数棋谜   妈！我会棋谜式乘法啦！
Mom! I Learn Multiplication Using Math-Chess-Puzzles Connection!

Student's Name _____ Date _____

2007 - 2018 © Frank Ho, Amanda Ho, All rights reserved.   www.homathchess.com

Learning multiplication with multi-concept and multi-format

c	9	8	7
b	2	6	6
a	3	4	5
	1	2	3

The original square is at b2 = □.

Mom! I Learn Multiplication Using Math-Chess-Puzzles Connection!

Student's Name _____ Date _____

2007 - 2018 © Frank Ho, Amanda Ho, All rights reserved.   www.homathchess.com

Learning multiplication with multi-concept and multi-format

c	9	8	7
b	2	6	6
a	3	4	5
	1	2	3

The original square is at b2 = □.

2007 - 2018 © Frank Ho, Amanda Ho, All rights reserved.　　www.homathchess.com

Learning multiplication with multi-concept and multi-format

c	9	8	7
b	2	6	6
a	3	4	5
	1	2	3

The original square is at b2 = □.

Student's Name _____ Date _____

2007 - 2018 © Frank Ho, Amanda Ho, All rights reserved. www.homathchess.com

Learning multiplication with multi-concept and multi-format

c	9	8	7
b	2	6	6
a	3	4	5
	1	2	3

The original square is at b2 = □.

$$\square \times \triangle = \bigcirc, \quad \bigcirc = \triangle \times \square, \quad \square \overline{)\dfrac{\triangle}{\bigcirc}}, \quad \triangle \overline{)\dfrac{\square}{\bigcirc}}, \quad \square \dfrac{\bigcirc}{\triangle},$$

$$\triangle \dfrac{\bigcirc}{\square}$$

$$\square = \bigcirc \times \dfrac{1}{\triangle}, \quad \triangle = \bigcirc \times \dfrac{1}{\square}, \quad \dfrac{1}{\square} = \dfrac{\triangle}{\bigcirc}, \quad \dfrac{1}{\triangle} = \dfrac{\square}{\bigcirc}, \quad \dfrac{\bigcirc}{\triangle} = \dfrac{\square}{1},$$

$$\dfrac{\bigcirc}{\square} = \dfrac{\triangle}{1}$$

$$\square \times \triangle = \bigcirc, \quad \bigcirc = \triangle \times \square, \quad \square \overline{)\dfrac{\triangle}{\bigcirc}}, \quad \triangle \overline{)\dfrac{\square}{\bigcirc}}, \quad \square \dfrac{\bigcirc}{\triangle},$$

$$\triangle \dfrac{\bigcirc}{\square}$$

$$\square = \bigcirc \times \dfrac{1}{\triangle}, \quad \triangle = \bigcirc \times \dfrac{1}{\square}, \quad \dfrac{1}{\square} = \dfrac{\triangle}{\bigcirc}, \quad \dfrac{1}{\triangle} = \dfrac{\square}{\bigcirc}, \quad \dfrac{\bigcirc}{\triangle} = \dfrac{\square}{1},$$

$$\dfrac{\bigcirc}{\square} = \dfrac{\triangle}{1}$$

Ho Math Chess   何数棋谜   妈！我会棋谜式乘法啦！
Mom! I Learn Multiplication Using Math-Chess-Puzzles Connection!
Student's Name _____ Date _____
2007 - 2018 © Frank Ho, Amanda Ho, All rights reserved.   www.homathchess.com

Learning multiplication with multi-concept and multi-format

c	9	8	7
b	2	6	6
a	3	4	5
	1	2	3

The original square is at b2 = □.

$$\square \times \triangle = \bigcirc, \quad \bigcirc = \triangle \times \square, \quad \square\overline{)\bigcirc}, \quad \triangle\overline{)\bigcirc}, \quad \square\frac{\bigcirc}{\triangle},$$

$$\triangle\frac{\bigcirc}{\square}$$

$$\square = \bigcirc \times \frac{1}{\triangle}, \quad \triangle = \bigcirc \times \frac{1}{\square}, \quad \frac{1}{\square} = \frac{\triangle}{\bigcirc}, \quad \frac{1}{\triangle} = \frac{\square}{\bigcirc}, \quad \frac{\bigcirc}{\triangle} = \frac{\square}{1},$$

$$\frac{\bigcirc}{\square} = \frac{\triangle}{1}$$

Ho Math Chess 何数棋谜 妈！我会棋谜式乘法啦！

Mom! I Learn Multiplication Using Math-Chess-Puzzles Connection!

Student's Name _____ Date _____

2007 - 2018 © Frank Ho, Amanda Ho, All rights reserved.   www.homathchess.com

Learning multiplication with multi-concept and multi-format

c	9	8	7
b	2	7	6
a	3	4	5
	1	2	3

The original square is at b2 = □.

$\square \times \underset{\longleftrightarrow}{\updownarrow} = \_\_ \times \_\_ = \_\_$ $\square \times \diagdown\!\!\!\!\diagup = \_\_ \times \_\_ = \_\_$

$\square \times \underset{\longleftrightarrow}{\updownarrow} = \_\_ \times \_\_ = \_\_$ $\square \times \diagdown\!\!\!\!\diagup = \_\_ \times \_\_ = \_\_$

$\square \times \underset{\longleftrightarrow}{\updownarrow} = \_\_ \times \_\_ = \_\_$ $\square \times \diagdown\!\!\!\!\diagup = \_\_ \times \_\_ = \_\_$

$\square \times \underset{\longleftrightarrow}{\updownarrow} = \_\_ \times \_\_ = \_\_$ $\square \times \diagdown\!\!\!\!\diagup = \_\_ \times \_\_ = \_\_$

$\square \times \triangle = \bigcirc, \quad \bigcirc = \triangle \times \square, \quad \square\sqrt{\dfrac{\triangle}{\bigcirc}}, \quad \triangle\sqrt{\dfrac{\square}{\bigcirc}}, \quad \square\dfrac{\bigcirc}{\triangle},$

$\triangle\dfrac{\bigcirc}{\square}$

$\square = \bigcirc \times \dfrac{1}{\triangle}, \quad \triangle = \bigcirc \times \dfrac{1}{\square}, \quad \dfrac{1}{\square} = \dfrac{\triangle}{\bigcirc}, \quad \dfrac{1}{\triangle} = \dfrac{\square}{\bigcirc}, \quad \dfrac{\bigcirc}{\triangle} = \dfrac{\square}{1},$

$\dfrac{\bigcirc}{\square} = \dfrac{\triangle}{1}$

254

Student's Name _____ Date _____

2007 - 2018 © Frank Ho, Amanda Ho, All rights reserved.   www.homathchess.com

Learning multiplication with multi-concept and multi-format

c	9	8	7
b	2	7	6
a	3	4	5
	1	2	3

The original square is at b2 = □.

Mom! I Learn Multiplication Using Math-Chess-Puzzles Connection!

Student's Name _____ Date _____

2007 - 2018 © Frank Ho, Amanda Ho, All rights reserved.   www.homathchess.com

Learning multiplication with multi-concept and multi-format

c	9	8	7
b	2	7	6
a	3	4	5
	1	2	3

The original square is at b2 = □.

$$\square \times \triangle = \bigcirc, \quad \bigcirc = \triangle \times \square, \quad \square\overline{)\bigcirc}, \quad \triangle\overline{)\bigcirc}, \quad \square\,\dfrac{\bigcirc}{\triangle}, \quad \triangle\,\dfrac{\bigcirc}{\square}$$

$$\square = \bigcirc \times \dfrac{1}{\triangle}, \quad \triangle = \bigcirc \times \dfrac{1}{\square}, \quad \dfrac{1}{\square} = \dfrac{\triangle}{\bigcirc}, \quad \dfrac{1}{\triangle} = \dfrac{\square}{\bigcirc}, \quad \dfrac{\bigcirc}{\triangle} = \dfrac{\square}{1}$$

$$\dfrac{\bigcirc}{\square} = \dfrac{\triangle}{1}$$

$$\square \times \triangle = \bigcirc, \quad \bigcirc = \triangle \times \square, \quad \square\overline{)\bigcirc}, \quad \triangle\overline{)\bigcirc}, \quad \square\,\dfrac{\bigcirc}{\triangle}, \quad \triangle\,\dfrac{\bigcirc}{\square}$$

$$\square = \bigcirc \times \dfrac{1}{\triangle}, \quad \triangle = \bigcirc \times \dfrac{1}{\square}, \quad \dfrac{1}{\square} = \dfrac{\triangle}{\bigcirc}, \quad \dfrac{1}{\triangle} = \dfrac{\square}{\bigcirc}, \quad \dfrac{\bigcirc}{\triangle} = \dfrac{\square}{1}$$

$$\dfrac{\bigcirc}{\square} = \dfrac{\triangle}{1}$$

2007 - 2018 © Frank Ho, Amanda Ho, All rights reserved.   www.homathchess.com

Learning multiplication with multi-concept and multi-format

c	9	8	7
b	2	7	6
a	3	4	5
	1	2	3

The original square is at b2 = □.

Ho Math Chess   何数棋谜   妈！我会棋谜式乘法啦！
Mom! I Learn Multiplication Using Math-Chess-Puzzles Connection!

Student's Name _____ Date _____

2007 - 2018 © Frank Ho, Amanda Ho, All rights reserved.   www.homathchess.com

Learning multiplication with multi-concept and multi-format

c	9	8	7
b	2	7	6
a	3	4	5
	1	2	3

The original square is at b2 = □.

$$\square \times \triangle = \bigcirc, \quad \bigcirc = \triangle \times \square, \quad \square \overline{)\dfrac{\triangle}{\bigcirc}}, \quad \triangle \overline{)\dfrac{\square}{\bigcirc}}, \quad \square \dfrac{\bigcirc}{\triangle},$$

$$\triangle \overline{)\dfrac{\bigcirc}{\square}}$$

$$\square = \bigcirc \times \dfrac{1}{\triangle}, \quad \triangle = \bigcirc \times \dfrac{1}{\square}, \quad \dfrac{1}{\square} = \dfrac{\triangle}{\bigcirc}, \quad \dfrac{1}{\triangle} = \dfrac{\square}{\bigcirc}, \quad \dfrac{\bigcirc}{\triangle} = \dfrac{\square}{1},$$

$$\dfrac{\bigcirc}{\square} = \dfrac{\triangle}{1}$$

2007 - 2018 © Frank Ho, Amanda Ho, All rights reserved.   www.homathchess.com

Learning multiplication with multi-concept and multi-format

c	9	8	7
b	2	8	6
a	3	4	5
	1	2	3

The original square is at b2 = □.

Ho Math Chess　　何数棋谜　妈！我会棋谜式乘法啦！
Mom! I Learn Multiplication Using Math-Chess-Puzzles Connection!

Student's Name _____ Date _____

2007 - 2018 © Frank Ho, Amanda Ho, All rights reserved.　　www.homathchess.com

Learning multiplication with multi-concept and multi-format

c	9	8	7
b	2	8	6
a	3	4	5
	1	2	3

The original square is at b2 = □.

$$\square \times \triangle = \bigcirc, \quad \bigcirc = \triangle \times \square, \quad \square \overline{)\dfrac{\triangle}{\bigcirc}}, \quad \triangle \overline{)\dfrac{\square}{\bigcirc}}, \quad \dfrac{\square \,)\,\bigcirc}{\triangle},$$

$$\triangle \,)\dfrac{\bigcirc}{\square}$$

$$\square = \bigcirc \times \dfrac{1}{\triangle}, \quad \triangle = \bigcirc \times \dfrac{1}{\square}, \quad \dfrac{1}{\square} = \dfrac{\triangle}{\bigcirc}, \quad \dfrac{1}{\triangle} = \dfrac{\square}{\bigcirc}, \quad \dfrac{\bigcirc}{\triangle} = \dfrac{\square}{1},$$

$$\dfrac{\bigcirc}{\square} = \dfrac{\triangle}{1}$$

$$\square \times \triangle = \bigcirc, \quad \bigcirc = \triangle \times \square, \quad \square \overline{)\dfrac{\triangle}{\bigcirc}}, \quad \triangle \overline{)\dfrac{\square}{\bigcirc}}, \quad \dfrac{\square \,)\,\bigcirc}{\triangle},$$

$$\triangle \,)\dfrac{\bigcirc}{\square}$$

$$\square = \bigcirc \times \dfrac{1}{\triangle}, \quad \triangle = \bigcirc \times \dfrac{1}{\square}, \quad \dfrac{1}{\square} = \dfrac{\triangle}{\bigcirc}, \quad \dfrac{1}{\triangle} = \dfrac{\square}{\bigcirc}, \quad \dfrac{\bigcirc}{\triangle} = \dfrac{\square}{1},$$

$$\dfrac{\bigcirc}{\square} = \dfrac{\triangle}{1}$$

Ho Math Chess   何数棋谜   妈!我会棋谜式乘法啦!
Mom! I Learn Multiplication Using Math-Chess-Puzzles Connection!

Student's Name _____ Date _____

2007 - 2018 © Frank Ho, Amanda Ho, All rights reserved.   www.homathchess.com

Learning multiplication with multi-concept and multi-format

c	9	8	7
b	2	8	6
a	3	4	5
	1	2	3

The original square is at b2 = $\square$.

Student's Name _____ Date _____

2007 - 2018 © Frank Ho, Amanda Ho, All rights reserved.   www.homathchess.com

Learning multiplication with multi-concept and multi-format

c	9	8	7
b	2	8	6
a	3	4	5
	1	2	3

The original square is at b2 = □.

$$\square \times \triangle = \bigcirc, \quad \bigcirc = \triangle \times \square, \quad \square\,\overline{)\,\tfrac{\triangle}{\bigcirc}}\,, \quad \triangle\,\overline{)\,\tfrac{\square}{\bigcirc}}\,, \quad \square : \tfrac{\bigcirc}{\triangle}\,, \quad \triangle : \tfrac{\bigcirc}{\square}\,,$$

$$\square = \bigcirc \times \dfrac{1}{\triangle}, \quad \triangle = \bigcirc \times \dfrac{1}{\square}, \quad \dfrac{1}{\square} = \dfrac{\triangle}{\bigcirc}, \quad \dfrac{1}{\triangle} = \dfrac{\square}{\bigcirc}, \quad \dfrac{\bigcirc}{\triangle} = \dfrac{\square}{1}, \quad \dfrac{\bigcirc}{\square} = \dfrac{\triangle}{1},$$

$$\dfrac{\bigcirc}{\square} = \dfrac{\triangle}{1}$$

$$\square \times \triangle = \bigcirc, \quad \bigcirc = \triangle \times \square, \quad \square\,\overline{)\,\tfrac{\triangle}{\bigcirc}}\,, \quad \triangle\,\overline{)\,\tfrac{\square}{\bigcirc}}\,, \quad \square : \tfrac{\bigcirc}{\triangle}\,, \quad \triangle : \tfrac{\bigcirc}{\square}\,,$$

$$\square = \bigcirc \times \dfrac{1}{\triangle}, \quad \triangle = \bigcirc \times \dfrac{1}{\square}, \quad \dfrac{1}{\square} = \dfrac{\triangle}{\bigcirc}, \quad \dfrac{1}{\triangle} = \dfrac{\square}{\bigcirc}, \quad \dfrac{\bigcirc}{\triangle} = \dfrac{\square}{1}, \quad \dfrac{\bigcirc}{\square} = \dfrac{\triangle}{1},$$

$$\dfrac{\bigcirc}{\square} = \dfrac{\triangle}{1}$$

Ho Math Chess 何数棋谜 妈！我会棋谜式乘法啦！

Mom! I Learn Multiplication Using Math-Chess-Puzzles Connection!

Student's Name _____ Date _____,_____

2007 - 2018 © Frank Ho, Amanda Ho, All rights reserved. www.homathchess.com

Learning multiplication with multi-concept and multi-format

c	9	8	7
b	2	8	6
a	3	4	5
	1	2	3

The original square is at b2 = □.

$$\square \times \triangle = \bigcirc, \quad \bigcirc = \triangle \times \square, \quad \square \overline{)\dfrac{\triangle}{\bigcirc}}, \quad \triangle \overline{)\dfrac{\square}{\bigcirc}}, \quad \dfrac{\square \, \, \bigcirc}{\triangle},$$

$$\triangle \, \, \dfrac{\bigcirc}{\square}$$

$$\square = \bigcirc \times \dfrac{1}{\triangle}, \quad \triangle = \bigcirc \times \dfrac{1}{\square}, \quad \dfrac{1}{\square} = \dfrac{\triangle}{\bigcirc}, \quad \dfrac{1}{\triangle} = \dfrac{\square}{\bigcirc}, \quad \dfrac{\bigcirc}{\triangle} = \dfrac{\square}{1},$$

$$\dfrac{\bigcirc}{\square} = \dfrac{\triangle}{1}$$

2007 - 2018 © Frank Ho, Amanda Ho, All rights reserved.   www.homathchess.com

Learning multiplication with multi-concept and multi-format

c	9	8	7
b	2	9	6
a	3	4	5
	1	2	3

The original square is at b2 = □.

Ho Math Chess 何数棋谜 妈!我会棋谜式乘法啦!
Mom! I Learn Multiplication Using Math-Chess-Puzzles Connection!

Student's Name _____ Date _____

2007 - 2018 © Frank Ho, Amanda Ho, All rights reserved.   www.homathchess.com

Learning multiplication with multi-concept and multi-format

c	9	8	7
b	2	9	6
a	3	4	5
	1	2	3

The original square is at **b2** = □.

$$\square \times \triangle = \bigcirc, \quad \bigcirc = \triangle \times \square, \quad \square\overline{)\dfrac{\triangle}{\bigcirc}}, \quad \triangle\overline{)\dfrac{\square}{\bigcirc}}, \quad \square\dfrac{\bigcirc}{\triangle},$$

$$\triangle\dfrac{\bigcirc}{\square}$$

$$\square = \bigcirc \times \dfrac{1}{\triangle}, \quad \triangle = \bigcirc \times \dfrac{1}{\square}, \quad \dfrac{1}{\square} = \dfrac{\triangle}{\bigcirc}, \quad \dfrac{1}{\triangle} = \dfrac{\square}{\bigcirc}, \quad \dfrac{\bigcirc}{\triangle} = \dfrac{\square}{1},$$

$$\dfrac{\bigcirc}{\square} = \dfrac{\triangle}{1}$$

$$\square \times \triangle = \bigcirc, \quad \bigcirc = \triangle \times \square, \quad \square\overline{)\dfrac{\triangle}{\bigcirc}}, \quad \triangle\overline{)\dfrac{\square}{\bigcirc}}, \quad \square\dfrac{\bigcirc}{\triangle},$$

$$\triangle\dfrac{\bigcirc}{\square}$$

$$\square = \bigcirc \times \dfrac{1}{\triangle}, \quad \triangle = \bigcirc \times \dfrac{1}{\square}, \quad \dfrac{1}{\square} = \dfrac{\triangle}{\bigcirc}, \quad \dfrac{1}{\triangle} = \dfrac{\square}{\bigcirc}, \quad \dfrac{\bigcirc}{\triangle} = \dfrac{\square}{1},$$

$$\dfrac{\bigcirc}{\square} = \dfrac{\triangle}{1}$$

2007 - 2018 © Frank Ho, Amanda Ho, All rights reserved.   www.homathchess.com

Learning multiplication with multi-concept and multi-format

c	9	8	7
b	2	9	6
a	3	4	5
	1	2	3

The original squ
are is at b2 = □.

$$\square \times \triangle = \bigcirc , \quad \bigcirc = \triangle \times \square , \quad \square \overline{)\dfrac{\triangle}{\bigcirc}} , \quad \triangle \overline{)\dfrac{\square}{\bigcirc}} , \quad \dfrac{\square}{\triangle}\bigcirc ,$$

$$\triangle \overline{)\dfrac{\bigcirc}{\square}}$$

$$\square = \bigcirc \times \dfrac{1}{\triangle} , \quad \triangle = \bigcirc \times \dfrac{1}{\square} , \quad \dfrac{1}{\square} = \dfrac{\triangle}{\bigcirc} , \quad \dfrac{1}{\triangle} = \dfrac{\square}{\bigcirc} , \quad \dfrac{\bigcirc}{\triangle} = \dfrac{\square}{1} ,$$

$$\dfrac{\bigcirc}{\square} = \dfrac{\triangle}{1}$$

$$\square \times \triangle = \bigcirc , \quad \bigcirc = \triangle \times \square , \quad \square \overline{)\dfrac{\triangle}{\bigcirc}} , \quad \triangle \overline{)\dfrac{\square}{\bigcirc}} , \quad \dfrac{\square}{\triangle}\bigcirc ,$$

$$\triangle \overline{)\dfrac{\bigcirc}{\square}}$$

$$\square = \bigcirc \times \dfrac{1}{\triangle} , \quad \triangle = \bigcirc \times \dfrac{1}{\square} , \quad \dfrac{1}{\square} = \dfrac{\triangle}{\bigcirc} , \quad \dfrac{1}{\triangle} = \dfrac{\square}{\bigcirc} , \quad \dfrac{\bigcirc}{\triangle} = \dfrac{\square}{1} ,$$

$$\dfrac{\bigcirc}{\square} = \dfrac{\triangle}{1}$$

Student's Name _____ Date _____

2007 - 2018 © Frank Ho, Amanda Ho, All rights reserved.   www.homathchess.com

Learning multiplication with multi-concept and multi-format

c	9	8	7
b	2	9	6
a	3	4	5
	1	2	3

The original square is at b2 = □.

$$\square \times \triangle = \bigcirc, \quad \bigcirc = \triangle \times \square, \quad \square\overline{)\dfrac{\triangle}{\bigcirc}}, \quad \triangle\overline{)\dfrac{\square}{\bigcirc}}, \quad \square \dfrac{)\bigcirc}{\triangle},$$

$$\triangle \dfrac{)\bigcirc}{\square}$$

$$\square = \bigcirc \times \dfrac{1}{\triangle}, \quad \triangle = \bigcirc \times \dfrac{1}{\square}, \quad \dfrac{1}{\square} = \dfrac{\triangle}{\bigcirc}, \quad \dfrac{1}{\triangle} = \dfrac{\square}{\bigcirc}, \quad \dfrac{\bigcirc}{\triangle} = \dfrac{\square}{1},$$

$$\dfrac{\bigcirc}{\square} = \dfrac{\triangle}{1}$$

$$\square \times \triangle = \bigcirc, \quad \bigcirc = \triangle \times \square, \quad \square\overline{)\dfrac{\triangle}{\bigcirc}}, \quad \triangle\overline{)\dfrac{\square}{\bigcirc}}, \quad \square \dfrac{)\bigcirc}{\triangle},$$

$$\triangle \dfrac{)\bigcirc}{\square}$$

$$\square = \bigcirc \times \dfrac{1}{\triangle}, \quad \triangle = \bigcirc \times \dfrac{1}{\square}, \quad \dfrac{1}{\square} = \dfrac{\triangle}{\bigcirc}, \quad \dfrac{1}{\triangle} = \dfrac{\square}{\bigcirc}, \quad \dfrac{\bigcirc}{\triangle} = \dfrac{\square}{1},$$

$$\dfrac{\bigcirc}{\square} = \dfrac{\triangle}{1}$$

2007 - 2018 © Frank Ho, Amanda Ho, All rights reserved.   www.homathchess.com

Learning multiplication with multi-concept and multi-format

c	9	8	7
b	2	9	6
a	3	4	5
	1	2	3

The original square is at b2 = □.

$$\square \times \triangle = \bigcirc, \quad \bigcirc = \triangle \times \square, \quad \square\sqrt{\dfrac{\triangle}{\bigcirc}}, \quad \triangle\sqrt{\dfrac{\square}{\bigcirc}}, \quad \square\dfrac{1\bigcirc}{\triangle},$$

$$\triangle\dfrac{1\bigcirc}{\square}$$

$$\square = \bigcirc \times \dfrac{1}{\triangle}, \quad \triangle = \bigcirc \times \dfrac{1}{\square}, \quad \dfrac{1}{\square} = \dfrac{\triangle}{\bigcirc}, \quad \dfrac{1}{\triangle} = \dfrac{\square}{\bigcirc}, \quad \dfrac{\bigcirc}{\triangle} = \dfrac{\square}{1},$$

$$\dfrac{\bigcirc}{\square} = \dfrac{\triangle}{1}$$

Ho Math Chess   何数棋谜   妈!我会棋谜式乘法啦!
Mom! I Learn Multiplication Using Math-Chess-Puzzles Connection!

Student's Name _____ Date _____

2007 - 2018 © Frank Ho, Amanda Ho, All rights reserved.   www.homathchess.com

Intelligent worksheets of students directed multiplication, addition and subtraction

3	1	2	3
2	4	5	6
1	7	8	9
	a	b	c

You are at b2 = ☐.

3	10	30	20
2	15	40	25
1	5	45	35
	d	e	f

.

☐ × ⬌ + or — (circle one) _____ = (shaded square on the right)

_____ × _____ + or — (circle one) _____ = _____

3			
2			x
1			
	d	e	f

☐ × ⬌ + or — (circle one) _____ = (shaded square on the right)

_____ × _____ + or — (circle one) _____ = _____

3			
2			
1		x	
	d	e	f

☐ × ⬌ + or — (circle one) _____ = (shaded square on the right)

_5_ × _6_ + or — (circle one) _5_ = _____

☐ × ⬌ + or — (circle one) _____ = (shaded square on the right)

_5_ × ___ + or — (circle one) _5_ = _____

Ho Math Chess   何数棋谜   妈！我会棋谜式乘法啦！
Mom! I Learn Multiplication Using Math-Chess-Puzzles Connection!

Student's Name _____ Date _____

2007 - 2018 © Frank Ho, Amanda Ho, All rights reserved.   www.homathchess.com

Intelligent worksheets of students directed multiplication, addition and subtraction

3	1	2	3
2	4	5	6
1	7	8	9
	a	b	c

You are at b2 = ☐.

3	10	30	20
2	15	40	25
1	5	45	35
	d	e	f

☐ × ✥ + or − (circle one) _____ = (shaded square on the right)

3			
2	✗		
1			
	d	e	f

_____ × ____ + or − (circle one) _____ = _____

☐ × ✥ + or − (circle one) _____ = (shaded square on the right)

3		✗	
2			
1			
	d	e	f

_____ × ____ + or − (circle one) _____ = _____

Ho Math Chess   何数棋谜   妈！我会棋谜式乘法啦！
Mom! I Learn Multiplication Using Math-Chess-Puzzles Connection!

Student's Name _____ Date _____

2007 - 2018 © Frank Ho, Amanda Ho, All rights reserved.   www.homathchess.com

Intelligent worksheets of students directed multiplication, addition and subtraction

3	1	2	3
2	4	5	6
1	7	8	9
	a	b	c

You are at b2 = ☐ .

3	10	30	20
2	15	40	25
1	5	45	35
	d	e	f

☐ × ✕ ✛ or ━ (circle one) _____ = (shaded square on the right)

_____ × _____ ✛ or ━ (circle one) _____ = _____

3	x		
2			
1			
	d	e	f

☐ × ✕ ✛ or ━ (circle one) _____ = (shaded square on the right)

_____ × _____ ✛ or ━ (circle one) _____ = _____

3			x
2			
1			
	d	e	f

2007 - 2018 © Frank Ho, Amanda Ho, All rights reserved.　www.homathchess.com

Intelligent worksheets of students directed multiplication, addition and subtraction

3	1	2	3
2	4	5	6
1	7	8	9
	a	b	c

You are at b2 = ☐ .

3	10	30	20
2	15	40	25
1	5	45	35
	d	e	f

☐ × ✕ **+** or **–** (circle one) _____ = (shaded square on the right)

_____ × _____ **+** or **–** (circle one) _____ = _____

3			
2			
1			▨
	d	e	f

☐ × ✕ **+** or **–** (circle one) _____ = (shaded square on the right)

_____ × _____ **+** or **–** (circle one) _____ = _____

3			
2			
1	▨		
	d	e	f

Ho Math Chess   何数棋谜   妈!我会棋谜式乘法啦!

Mom! I Learn Multiplication Using Math-Chess-Puzzles Connection!

Student's Name _____ Date _____

2007 - 2018 © Frank Ho, Amanda Ho, All rights reserved.   www.homathchess.com

Intelligent worksheets of students directed multiplication, addition and subtraction

3	1	2	3
2	4	6	6
1	7	8	9
	a	b	c

You are at b2 = ☐.

3	54	12	20
2	48	30	18
1	36	24	42
	d	e	f

☐ × ⬌ + or — (circle one) _____ = (shaded square on the right)

_____ × _____ + or — (circle one) _____ = _____

3			
2			x
1			
	d	e	f

☐ × ⬌ + or — (circle one) _____ = (shaded square on the right)

_____ × _____ + or — (circle one) _____ = _____

3			
2			
1		x	
	d	e	f

Ho Math Chess 何数棋谜 妈！我会棋谜式乘法啦！
Mom! I Learn Multiplication Using Math-Chess-Puzzles Connection!

Student's Name _____ Date _____

2007 - 2018 © Frank Ho, Amanda Ho, All rights reserved.   www.homathchess.com

Intelligent worksheets of students directed multiplication, addition and subtraction

3	1	2	3
2	4	6	6
1	7	8	9
	a	b	c

You are at b2 = ☐ .

3	54	12	20
2	48	30	18
1	36	24	42
	d	e	f

☐ × ⬍ + or — (circle one) _____ = (shaded square on the right)

3			
2	X		
1			
	d	e	f

_____ × _____ + or — (circle one) _____ = _____

☐ × ⬍ + or — (circle one) _____ = (shaded square on the right)

3		X	
2			
1			
	d	e	f

_____ × _____ + or — (circle one) _____ = _____

Student's Name _____ Date _____

2007 - 2018 © Frank Ho, Amanda Ho, All rights reserved.   www.homathchess.com

Intelligent worksheets of students directed multiplication, addition and subtraction

3	1	2	3
2	4	6	6
1	7	8	9
	a	b	c

You are at b2 = ☐.

3	54	12	20
2	48	30	18
1	36	24	42
	d	e	f

☐ × ╳ ✛ or ━ (circle one) _____ = (shaded square on the right)

_____ × _____ ✛ or ━ (circle one) _____ = _____

3	▨		
2			
1			
	d	e	f

☐ × ╳ ✛ or ━ (circle one) _____ = (shaded square on the right)

_____ × _____ ✛ or ━ (circle one) _____ = _____

3			▨
2			
1			
	d	e	f

Ho Math Chess   何数棋谜   妈！我会棋谜式乘法啦！

Mom! I Learn Multiplication Using Math-Chess-Puzzles Connection!

Student's Name _____ Date _____

2007 - 2018 © Frank Ho, Amanda Ho, All rights reserved.   www.homathchess.com

Intelligent worksheets of students directed multiplication, addition and subtraction

3	1	2	3
2	4	6	6
1	7	8	9
	a	b	c

You are at b2 = ☐ .

3	54	12	20
2	48	30	18
1	36	24	42
	d	e	f

☐ ✕ ╳ ➕ or ➖ (circle one) _____ = (shaded square on the right)

_____ ✕ _____ ➕ or ➖ (circle one) _____ = _____

3			
2			
1			▨
	d	e	f

☐ ✕ ╳ ➕ or ➖ (circle one) _____ = (shaded square on the right)

_____ ✕ _____ ➕ or ➖ (circle one) _____ = _____

3			
2			
1	▨		
	d	e	f

Student's Name _____ Date _____

2007 - 2018 © Frank Ho, Amanda Ho, All rights reserved.   www.homathchess.com

Intelligent worksheets of students directed multiplication, addition and subtraction

3	1	2	3
2	4	7	6
1	7	8	9
	a	b	c

You are at b2 = ☐.

3	31	30	63
2	15	28	56
1	49	42	35
	d	e	f

☐ × ✛ + or — (circle one) _____ = (shaded square on the right)

_____ × ____ + or — (circle one) _____ = _____

3			
2			X
1			
	d	e	f

☐ × ✛ + or — (circle one) _____ = (shaded square on the right)

_____ × ____ + or — (circle one) _____ = _____

3			
2			
1		X	
	d	e	f

Ho Math Chess   何数棋谜   妈！我会棋谜式乘法啦！
Mom! I Learn Multiplication Using Math-Chess-Puzzles Connection!

Student's Name _____ Date _____

2007 - 2018 © Frank Ho, Amanda Ho, All rights reserved.   www.homathchess.com

Intelligent worksheets of students directed multiplication, addition and subtraction

3	1	2	3
2	4	7	6
1	7	8	9
	a	b	c

You are at b2 = ☐ .

3	31	30	63
2	15	28	56
1	49	42	35
	d	e	f

☐ × ⤡ + or — (circle one) _____ = (shaded square on the right)

_____ × ____ + or — (circle one) _____ = _____

3			
2	x		
1			
	d	e	f

☐ × ⤡ + or — (circle one) _____ = (shaded square on the right)

_____ × ____ + or — (circle one) _____ = _____

3		x	
2			
1			
	d	e	f

2007 - 2018 © Frank Ho, Amanda Ho, All rights reserved.   www.homathchess.com

Intelligent worksheets of students directed multiplication, addition and subtraction

3	1	2	3
2	4	7	6
1	7	8	9
	a	b	c

You are at b2 = ☐.

3	31	30	63
2	15	28	56
1	49	42	35
	d	e	f

☐ × ✕ + or − (circle one) _____ = (shaded square on the right)

_____ × _____ + or − (circle one) _____ = _____

3	x		
2			
1			
	d	e	f

☐ × ✕ + or − (circle one) _____ = (shaded square on the right)

_____ × _____ + or − (circle one) _____ = _____

3			x
2			
1			
	d	e	f

2007 - 2018 © Frank Ho, Amanda Ho, All rights reserved.　　www.homathchess.com

Intelligent worksheets of students directed multiplication, addition and subtraction

3	1	2	3
2	4	7	6
1	7	8	9
	a	b	c

You are at b2 = ☐.

3	31	30	63
2	15	28	56
1	49	42	35
	d	e	f

☐ × ╳ ➕ or ➖ (circle one) _____ = (shaded square on the right)

_____ × _____ ➕ or ➖ (circle one) _____ = _____

3			
2			
1			X
	d	e	f

☐ × ╳ ➕ or ➖ (circle one) _____ = (shaded square on the right)

_____ × _____ ➕ or ➖ (circle one) _____ = _____

3			
2			
1	X		
	d	e	f

Student's Name _____ Date _____

2007 - 2018 © Frank Ho, Amanda Ho, All rights reserved.   www.homathchess.com

Intelligent worksheets of students directed multiplication, addition and subtraction

3	1	2	3
2	4	8	6
1	7	8	9
	a	b	c

You are at b2 = ☐.

3	48	56	24
2	64	40	72
1	24	32	16
	d	e	f

☐ × ⬌⬍ + or − (circle one) _____ = (shaded square on the right)

_____ × ____ + or − (circle one) _____ = _____

3			
2			X
1			
	d	e	f

☐ × ⬌⬍ + or − (circle one) _____ = (shaded square on the right)

_____ × ____ + or − (circle one) _____ = _____

3			
2			
1		X	
	d	e	f

Ho Math Chess　　何数棋谜　妈！我会棋谜式乘法啦！
Mom! I Learn Multiplication Using Math-Chess-Puzzles Connection!

Student's Name _____ Date _____

2007 - 2018 © Frank Ho, Amanda Ho, All rights reserved.　　www.homathchess.com

Intelligent worksheets of students directed multiplication, addition and subtraction

3	1	2	3
2	4	8	6
1	7	8	9
	a	b	c

You are at b2 = ☐.

3	48	56	24
2	64	40	72
1	24	32	16
	d	e	f

☐ × ⊕ + or − (circle one) _____ = (shaded square on the right)

_____ × _____ + or − (circle one) _____ = _____

3			
2	x		
1			
	d	e	f

☐ × ⊕ + or − (circle one) _____ = (shaded square on the right)

_____ × _____ + or − (circle one) _____ = _____

3		x	
2			
1			
	d	e	f

Mom! I Learn Multiplication Using Math-Chess-Puzzles Connection!

Student's Name _____ Date _____

2007 - 2018 © Frank Ho, Amanda Ho, All rights reserved.   www.homathchess.com

Intelligent worksheets of students directed multiplication, addition and subtraction

3	1	2	3
2	4	8	6
1	7	8	9
	a	b	c

You are at b2 = ☐.

3	48	56	24
2	64	40	72
1	24	32	16
	d	e	f

☐ × ╳ + or − (circle one) _____ = (shaded square on the right)

3	x		
2			
1			
	d	e	f

_____ × _____ + or − (circle one) _____ = _____

☐ × ╳ + or − (circle one) _____ = (shaded square on the right)

3			x
2			
1			
	d	e	f

_____ × _____ + or − (circle one) _____ = _____

Ho Math Chess   何数棋谜   妈！我会棋谜式乘法啦！
Mom! I Learn Multiplication Using Math-Chess-Puzzles Connection!

Student's Name _____ Date _____

2007 - 2018 © Frank Ho, Amanda Ho, All rights reserved.   www.homathchess.com

Intelligent worksheets of students directed multiplication, addition and subtraction

3	1	2	3
2	4	8	6
1	7	8	9
	a	b	c

You are at b2 = ☐ .

3	48	56	24
2	64	40	72
1	24	32	16
	d	e	f

☐ × ╳ + or — (circle one) _____ = (shaded square on the right)

3			
2			
1			x
	d	e	f

_____ × _____ + or — (circle one) _____ = _____

☐ × ╳ + or — (circle one) _____ = (shaded square on the right)

3			
2			
1	x		
	d	e	f

_____ × _____ + or — (circle one) _____ = _____

Ho Math Chess   何数棋谜   妈！我会棋谜式乘法啦！
Mom! I Learn Multiplication Using Math-Chess-Puzzles Connection!

Student's Name _____ Date _____

2007 - 2018 © Frank Ho, Amanda Ho, All rights reserved.   www.homathchess.com

Intelligent worksheets of students directed multiplication, addition and subtraction

3	1	2	3
2	4	9	6
1	7	8	9
	a	b	c

3	36	81	27
2	63	40	54
1	18	45	36
	d	e	f

You are at b2 = ☐ .

☐ × ⬍ + or − (circle one) _____ = (shaded square on the right)

_____ × ____ + or − (circle one) _____ = _____

3			
2			x
1			
	d	e	f

☐ × ⬍ + or − (circle one) _____ = (shaded square on the right)

_____ × ____ + or − (circle one) _____ = _____

3			
2			
1		x	
	d	e	f

Mom! I Learn Multiplication Using Math-Chess-Puzzles Connection!

Student's Name _____ Date _____

2007 - 2018 © Frank Ho, Amanda Ho, All rights reserved.   www.homathchess.com

Intelligent worksheets of students directed multiplication, addition and subtraction

3	1	2	3
2	4	9	6
1	7	8	9
	a	b	c

You are at b2 = ☐ .

3	36	81	27
2	63	40	54
1	18	45	36
	d	e	f

☐ × ✛ + or − (circle one) _____ = (shaded square on the right)

_____ × _____ + or − (circle one) _____ = _____

3			
2	x		
1			
	d	e	f

☐ × ✛ + or − (circle one) _____ = (shaded square on the right)

_____ × _____ + or − (circle one) _____ = _____

3		x	
2			
1			
	d	e	f

Student's Name _____ Date _____

2007 - 2018 © Frank Ho, Amanda Ho, All rights reserved.   www.homathchess.com

Intelligent worksheets of students directed multiplication, addition and subtraction

3	1	2	3
2	4	9	6
1	7	8	9
	a	b	c

You are at b2 = ☐ .

3	36	81	27
2	63	40	54
1	18	45	36
	d	e	f

☐ × ✕ + or — (circle one) _____ = (shaded square on the right)

_____ × _____ + or — (circle one) _____ = _____

3	X		
2			
1			
	d	e	f

☐ × ✕ + or — (circle one) _____ = (shaded square on the right)

_____ × _____ + or — (circle one) _____ = _____

3			X
2			
1			
	d	e	f

2007 - 2018 © Frank Ho, Amanda Ho, All rights reserved.   www.homathchess.com

Intelligent worksheets of students directed multiplication, addition and subtraction

3	1	2	3
2	4	9	6
1	7	8	9
	a	b	c

You are at b2 = ☐.

3	36	81	27
2	63	40	54
1	18	45	36
	d	e	f

☐ × ✕ + or − (circle one) _____ = (shaded square on the right)

_____ × _____ + or − (circle one) _____ = _____

3			
2			
1			x
	d	e	f

☐ × ✕ + or − (circle one) _____ = (shaded square on the right)

_____ × _____ + or − (circle one) _____ = _____

3			
2			
1	x		
	d	e	f

Ho Math Chess   何数棋谜   妈！我会棋谜式乘法啦！
Mom! I Learn Multiplication Using Math-Chess-Puzzles Connection!

Student's Name _____ Date _____

2007 - 2018 © Frank Ho, Amanda Ho, All rights reserved.   www.homathchess.com

Intelligent worksheets of students directed multiplication, addition and subtraction

3	1	2	3
2	4	4	6
1	7	8	9
	a	b	c

3	36	27	12
2	24	28	8
1	20	32	16
	d	e	f

You are at b2 = ☐.

☐ × ✥ + or − (circle one) _____ = (shaded square on the right)

_____ × ____ + or − (circle one) _____ = _____

3			
2			x
1			
	d	e	f

☐ × ⚓ + or − (circle one) _____ = (shaded square on the right)

_____ × ____ + or − (circle one) _____ = _____

3			
2			
1		x	
	d	e	f

Mom! I Learn Multiplication Using Math-Chess-Puzzles Connection!

Student's Name _____ Date _____

2007 - 2018 © Frank Ho, Amanda Ho, All rights reserved.   www.homathchess.com

Intelligent worksheets of students directed multiplication, addition and subtraction

3	1	2	3
2	4	4	6
1	7	8	9
	a	b	c

You are at b2 = ☐.

3	36	27	12
2	24	28	8
1	20	32	16
	d	e	f

☐ × ⬌ + or — (circle one) _____ = (shaded square on the right)

_____ × _____ + or — (circle one) _____ = _____

3			
2	x		
1			
	d	e	f

☐ × ⬌ + or — (circle one) _____ = (shaded square on the right)

_____ × _____ + or — (circle one) _____ = _____

3		x	
2			
1			
	d	e	f

Mom! I Learn Multiplication Using Math-Chess-Puzzles Connection!

Student's Name _____ Date _____

2007 - 2018 © Frank Ho, Amanda Ho, All rights reserved.   www.homathchess.com

Intelligent worksheets of students directed multiplication, addition and subtraction

3	1	2	3
2	4	4	6
1	7	8	9
	a	b	c

You are at b2 = ☐.

3	36	27	12
2	24	28	8
1	20	32	16
	d	e	f

☐ × ✕ **+** or **—** (circle one) _____ = (shaded square on the right)

_____ × _____ **+** or **—** (circle one) _____ = _____

3	x		
2			
1			
	d	e	f

☐ × ✕ **+** or **—** (circle one) _____ = (shaded square on the right)

_____ × _____ **+** or **—** (circle one) _____ = _____

3			x
2			
1			
	d	e	f

Student's Name _____ Date _____

2007 - 2018 © Frank Ho, Amanda Ho, All rights reserved.   www.homathchess.com

Intelligent worksheets of students directed multiplication, addition and subtraction

3	1	2	3
2	4	4	6
1	7	8	9
	a	b	c

You are at b2 = ☐.

3	36	27	12
2	24	28	8
1	20	32	16
	d	e	f

☐ × ✕ ➕ or ➖ (circle one) _____ = (shaded square on the right)

_____ × ____ ➕ or ➖ (circle one) _____ = _____

3			
2			
1			x
	d	e	f

☐ × ✕ ➕ or ➖ (circle one) _____ = (shaded square on the right)

_____ × ____ ➕ or ➖ (circle one) _____ = _____

3			
2			
1	x		
	d	e	f

Ho Math Chess　何数棋谜　妈！我会棋谜式乘法啦！

Mom! I Learn Multiplication Using Math-Chess-Puzzles Connection!

Student's Name _____ Date _____

2007 - 2018 © Frank Ho, Amanda Ho, All rights reserved.　www.homathchess.com

Intelligent worksheets of students directed multiplication, addition and subtraction

3	1	2	3
2	4	3	6
1	7	8	9
	a	b	c

You are at b2 = ☐.

3	6	27	12
2	15	24	21
1	12	9	18
	d	e	f

☐ × ⬍ + or − (circle one) _____ = (shaded square on the right)

_____ × ____ + or − (circle one) _____ = _____

3			
2			x
1			
	d	e	f

☐ × ⬍ + or − (circle one) _____ = (shaded square on the right)

_____ × ____ + or − (circle one) _____ = _____

3			
2			
1		x	
	d	e	f

Ho Math Chess   何数棋谜   妈！我会棋谜式乘法啦！
Mom! I Learn Multiplication Using Math-Chess-Puzzles Connection!

Student's Name _____ Date _____

2007 - 2018 © Frank Ho, Amanda Ho, All rights reserved.   www.homathchess.com

Intelligent worksheets of students directed multiplication, addition and subtraction

3	1	2	3
2	4	3	6
1	7	8	9
	a	b	c

You are at b2 = ☐.

3	6	27	12
2	15	24	21
1	12	9	18
	d	e	f

☐ × ⬌ + or − (circle one) _____ = (shaded square on the right)

_____ × ____ + or − (circle one) _____ = _____

☐ × ⬌ + or − (circle one) _____ = (shaded square on the right)

_____ × ____ + or − (circle one) _____ = _____

Ho Math Chess 何数棋谜 妈！我会棋谜式乘法啦！

Mom! I Learn Multiplication Using Math-Chess-Puzzles Connection!

Student's Name _____ Date _____

2007 - 2018 © Frank Ho, Amanda Ho, All rights reserved.   www.homathchess.com

Intelligent worksheets of students directed multiplication, addition and subtraction

3	1	2	3
2	4	3	6
1	7	8	9
	a	b	c

You are at b2 = ☐.

3	6	27	12
2	15	24	21
1	12	9	18
	d	e	f

☐ × ╳ + or − (circle one) _____ = (shaded square on the right)

_____ × _____ + or − (circle one) _____ = _____

3	x		
2			
1			
	d	e	f

☐ × ╳ + or − (circle one) _____ = (shaded square on the right)

_____ × ____ + or − (circle one) _____ = _____

3			x
2			
1			
	d	e	f

Ho Math Chess   何数棋谜   妈！我会棋谜式乘法啦！
Mom! I Learn Multiplication Using Math-Chess-Puzzles Connection!

Student's Name _____ Date _____

2007 - 2018 © Frank Ho, Amanda Ho, All rights reserved.   www.homathchess.com

Intelligent worksheets of students directed multiplication, addition and subtraction

3	1	2	3
2	4	3	6
1	7	8	9
	a	b	c

You are at b2 = ☐ .

3	6	27	12
2	15	24	21
1	12	9	18
	d	e	f

☐ × ✕ + or — (circle one) _____ = (shaded square on the right)

_____ × _____ + or — (circle one) _____ = _____

3			
2			
1			X
	d	e	f

☐ × ✕ + or — (circle one) _____ = (shaded square on the right)

_____ × _____ + or — (circle one) _____ = _____

3			
2			
1	X		
	d	e	f

2007 - 2018 © Frank Ho, Amanda Ho, All rights reserved.   www.homathchess.com

Mutiplication table

×	200	201	202	203	204	205	206	207	208	209
♟										
2										
♝										
4										
♜										
6										
7										
8										
♛										

2007 - 2018 © Frank Ho, Amanda Ho, All rights reserved.   www.homathchess.com

Multiplication table

✕	300	301	302	303	304	305	306	307	308	309
♙										
2										
♗										
4										
♖										
6										
7										
8										
♕										

Ho Math Chess 何数棋谜 妈！我会棋谜式乘法啦！
Mom! I Learn Multiplication Using Math-Chess-Puzzles Connection!
Student's Name _____ Date _____
2007 - 2018 © Frank Ho, Amanda Ho, All rights reserved.   www.homathchess.com

Multiplication table

✕	400	401	402	403	404	405	406	407	408	409
♟										
2										
♝										
4										
♜										
6										
7										
8										
♛										

Mom! I Learn Multiplication Using Math-Chess-Puzzles Connection!

Student's Name _____ Date _____

2007 - 2018 © Frank Ho, Amanda Ho, All rights reserved.   www.homathchess.com

Multiplication table

✕	500	501	502	503	504	505	506	507	508	509
♙										
2										
♗										
4										
♖										
6										
7										
8										
♕										

2007 - 2018 © Frank Ho, Amanda Ho, All rights reserved.   www.homathchess.com

Multiplication table

×	600	601	602	603	604	605	606	607	608	609
♙										
2										
♗										
4										
♖										
6										
7										
8										
♕										

Mom! I Learn Multiplication Using Math-Chess-Puzzles Connection!

Student's Name _____ Date _____

2007 - 2018 © Frank Ho, Amanda Ho, All rights reserved. www.homathchess.com

Multiplication table

×	700	701	702	703	704	705	706	707	708	709
♙										
2										
♝										
4										
♜										
6										
7										
8										
♛										

2007 - 2018 © Frank Ho, Amanda Ho, All rights reserved.　　www.homathchess.com

Multiplication table

×	800	801	802	803	804	805	806	807	808	809
♟										
2										
♝										
4										
♜										
6										
7										
8										
♛										

2007 - 2018 © Frank Ho, Amanda Ho, All rights reserved.　　www.homathchess.com

Multiplication table

×	900	901	902	903	904	905	906	907	908	909
♙										
2										
♗										
4										
♖										
6										
7										
8										
♕										

2007 - 2018 © Frank Ho, Amanda Ho, All rights reserved. www.homathchess.com

Multiplication table

×	11	12	13	14	15	16	17	18	19
♙									
2									
♗									
4									
♖									
6									
7									
8									
♛									

2007 - 2018 © Frank Ho, Amanda Ho, All rights reserved.　　www.homathchess.com

Multiplication table

×	111	121	131	141	151	161	171	181	191
♙									
2									
♗									
4									
♖									
6									
7									
8									
♕									

2007 - 2018 © Frank Ho, Amanda Ho, All rights reserved.   www.homathchess.com

Multiplication table

×	11	22	33	44	55	66	77	88	99
♙									
2									
♗									
4									
♖									
6									
7									
8									
♕									

2007 - 2018 © Frank Ho, Amanda Ho, All rights reserved.   www.homathchess.com

dd ✕ d **with carrying**

1	1	1	1	1
1 2	1 3	1 2	1 2	1 2
x 6	x 6	x 7	x 8	x ♖
7 2	☐☐	☐☐	☐☐	☐☐

1 3	1 3	1 3	1 3	1 3
x ♖	x 6	x 7	x 8	x ♕
☐☐	☐☐	☐☐	☐☐	☐☐☐

2	1	2	3	2
1 4	1 4	1 4	1 4	1 4
x 6	x 4	x 7	x 8	x 5
☐☐	☐☐	☐☐	☐☐	☐☐

1 5	1 5	1 5	1 5	1 5
x ♖	x 6	x 7	x 8	x 9
☐☐	☐☐	☐☐☐	☐☐☐	☐☐☐

Student's Name _____ Date _____

2007 - 2018 © Frank Ho, Amanda Ho, All rights reserved.   www.homathchess.com

dd ✕ d with carrying

```
  1            1            1            1            1
 2 2         2 3          2 2          2 2          2 2
x  6        x  6         x  7         x  8         x  ♖
1 3 2
```

```
 2 3         2 3          2 3          2 3          2 3
x  5        x  6         x  7         x  8         x  9
```

```
  2            1            2            3            2
 2 4         2 4          2 4          2 4          2 4
x  6        x  4         x  7         x  8         x  ♖
```

```
 2 5         2 5          2 5          2 5          2 5
x  5        x  6         x  7         x  8         x  ♕
```

Ho Math Chess   何数棋谜   妈！我会棋谜式乘法啦！
Mom! I Learn Multiplication Using Math-Chess-Puzzles Connection!
Student's Name _____ Date _____
2007 - 2018 © Frank Ho, Amanda Ho, All rights reserved.   www.homathchess.com

dd ✕ d with carrying

1	1	1	1	1
3 2	3 3	3 2	3 2	3 2
x 6	x 6	x 7	x 8	x 5

3 3	3 3	3 3	3 3	3 3
x ♜	x 6	x 7	x 8	x ♛

2	1	2	3	2
3 4	3 4	3 4	3 4	3 4
x 6	x 4	x 7	x 8	x ♜

3 5	3 5	3 5	3 5	3 5
x ♜	x 6	x 7	x 8	x ♛

2007 - 2018 © Frank Ho, Amanda Ho, All rights reserved.   www.homathchess.com

dd ✕ **d with carrying**

1	1	1	1	1
4 2	4 3	4 2	4 2	4 2
x 6	x 6	x 7	x 8	x ▦

4 3	4 3	4 3	4 3	4 3
x 5	x 6	x 7	x 8	x ♛

2	1	2	3	2
4 4	4 4	4 4	4 4	4 4
x 6	x 4	x 7	x 8	x 5

0.8

4 5	4 5	4 5	4 5	4 5
x ▦	x 6	x 7	x 8	x ♛

Mom! I Learn Multiplication Using Math-Chess-Puzzles Connection!

Student's Name _____ Date _____

2007 - 2018 © Frank Ho, Amanda Ho, All rights reserved.   www.homathchess.com

dd × d with carrying

¹52	¹53	¹52	¹52	¹52
x 6	x 6	x 7	x 8	x ♜

53	53	53	53	53
x ♜	x 6	x 7	x 8	x ♛

²54	¹54	²54	³54	²54
x 6	x 4	x 7	x 8	x 5

55	55	55	55	55
x ♜	x 6	x 7	x 8	x ♛

2007 - 2018 © Frank Ho, Amanda Ho, All rights reserved.   www.homathchess.com

dd ✕ d with carrying

1	1	1	1	1
6 2	6 3	6 2	6 2	6 2
x 6	x 6	x 7	x 8	x 5

6 3	6 3	6 3	6 3	6 3
x ♖	x 6	x 7	x 8	x ♕

2	1	2	3	2
6 4	6 4	6 4	6 4	6 4
x 6	x 4	x 7	x 8	x ♖

6 5	6 5	6 5	6 5	6 5
x ♖	x 6	x 7	x 8	x ♕

2007 - 2018 © Frank Ho, Amanda Ho, All rights reserved.   www.homathchess.com

dd ✕ d with carrying

1	1	1	1	1
7 2	7 3	7 2	7 2	7 2
x 6	x 6	x 7	x 8	x ♖

	1			
7 3	7 3	7 3	7 3	7 3
x ♖	x 6	x 7	x 8	x ♕

2	1	2	3	2
7 4	7 4	7 4	7 4	7 4
x 6	x 4	x 7	x 8	x 5

7 5	7 5	7 5	7 5	7 5
x ♖	x 6	x 7	x 8	x ♕

dd ✕ d with carrying

1	1	1	1	1
8 2	8 3	8 2	8 2	8 2
x 6	x 6	x 7	x 8	x ♖

8 3	8 3	8 3	8 3	8 3
x ♖	x 6	x 7	x 8	x ♕

2	1	2	3	2
8 4	8 4	8 4	8 4	8 4
x 6	x 4	x 7	x 8	x ♖

8 5	8 5	8 5	8 5	8 5
x ♖	x 6	x 7	x 8	x ♕

Mom! I Learn Multiplication Using Math-Chess-Puzzles Connection!

Student's Name _____ Date _____

2007 - 2018 © Frank Ho, Amanda Ho, All rights reserved.   www.homathchess.com

dd ✕ d with carrying

¹9 2	¹9 3	¹9 2	¹9 2	¹9 2
x 6	x 6	x 7	x 8	x ♖

9 3	9 3	9 3	9 3	9 3
x ♖	x 6	x 7	x 8	x ♕

²9 4	¹9 4	²9 4	³9 4	²9 4
x 6	x 4	x 7	x 8	x ♖

9 5	9 5	9 5	9 5	9 5
x ♖	x 6	x 7	x 8	x ♕

Ho Math Chess 何数棋谜 妈！我会棋谜式乘法啦！
Mom! I Learn Multiplication Using Math-Chess-Puzzles Connection!

Student's Name _____ Date _____

2007 - 2018 © Frank Ho, Amanda Ho, All rights reserved.   www.homathchess.com

Multiplication table

×	11	11	11	11	11	11	11	11
♙								
2								
♗								
4								
♖								
6								
7								
8								
♛								

317

Mom! I Learn Multiplication Using Math-Chess-Puzzles Connection!

Student's Name _____ Date _____

2007 - 2018 © Frank Ho, Amanda Ho, All rights reserved.　　www.homathchess.com

Multiplication table

×	22	22	22	22	22	22	22	22
♙								
2								
♗								
4								
♖								
6								
7								
8								
♛								

Mom! I Learn Multiplication Using Math-Chess-Puzzles Connection!

Student's Name _____ Date _____

2007 - 2018 © Frank Ho, Amanda Ho, All rights reserved.　　www.homathchess.com

Multiplication table

×	33	33	33	33	33	33	33	33
♙								
2								
♗								
4								
♖								
6								
7								
8								
♕								

Mom! I Learn Multiplication Using Math-Chess-Puzzles Connection!

Student's Name _____ Date _____

2007 - 2018 © Frank Ho, Amanda Ho, All rights reserved.　　www.homathchess.com

Multiplication table

×	44	44	44	44	44	44	44	44
♙								
2								
♗								
4								
♖								
6								
7								
8								
♕								

Mom! I Learn Multiplication Using Math-Chess-Puzzles Connection!

Student's Name _____ Date _____

2007 - 2018 © Frank Ho, Amanda Ho, All rights reserved.   www.homathchess.com

Multiplication table

×	55	55	55	55	55	55	55	55
♟								
2								
♝								
4								
♜								
6								
7								
8								
♛								

Ho Math Chess 何数棋谜 妈!我会棋谜式乘法啦!

Mom! I Learn Multiplication Using Math-Chess-Puzzles Connection!

Student's Name _____ Date _____

2007 - 2018 © Frank Ho, Amanda Ho, All rights reserved. www.homathchess.com

Multiplication table

×	66	66	66	66	66	66	66	66
♙								
2								
♗								
4								
♖								
6								
7								
8								
♕								

2007 - 2018 © Frank Ho, Amanda Ho, All rights reserved.   www.homathchess.com

Multiplication table

×	77	77	77	77	77	77	77	77
♙								
2								
♝								
4								
♜								
6								
7								
8								
♛								

Mom! I Learn Multiplication Using Math-Chess-Puzzles Connection!

Student's Name _____ Date _____

2007 - 2018 © Frank Ho, Amanda Ho, All rights reserved.　　www.homathchess.com

Multiplication table

×	88	88	88	88	88	88	88	88
♙								
2								
♗								
4								
♖								
6								
7								
8								
♛								

2007 - 2018 © Frank Ho, Amanda Ho, All rights reserved.   www.homathchess.com

Multiplication table

✕	99	99	99	99	99	99	99	99
♙								
2								
♗								
4								
♖								
6								
7								
8								
♕								

Ho Math Chess　何数棋谜　妈!我会棋谜式乘法啦!

Mom! I Learn Multiplication Using Math-Chess-Puzzles Connection!

Student's Name _____ Date _____

2007 - 2018 © Frank Ho, Amanda Ho, All rights reserved.　www.homathchess.com

ddd ✕ d with carrying

$$\begin{array}{r} 1\ 1\ 2 \\ \times\ \ \ 6 \\ \hline \end{array} \qquad \begin{array}{r} 1\ 1\ 2 \\ \times\ \ \ 9 \\ \hline \end{array} \qquad \begin{array}{r} 1\ 1\ 2 \\ \times\ \ \ 7 \\ \hline \end{array} \qquad \begin{array}{r} 1\ 1\ 2 \\ \times\ \ \ 8 \\ \hline \end{array} \qquad \begin{array}{r} 1\ 1\ 2 \\ \times\ \ ♛ \\ \hline \end{array}$$

$$\begin{array}{r} 1\ 1\ 3 \\ \times\ \ ♜ \\ \hline \end{array} \qquad \begin{array}{r} 1\ 1\ 3 \\ \times\ \ \ 6 \\ \hline \end{array} \qquad \begin{array}{r} 1\ 1\ 3 \\ \times\ \ \ 7 \\ \hline \end{array} \qquad \begin{array}{r} 1\ 1\ 3 \\ \times\ \ \ 8 \\ \hline \end{array} \qquad \begin{array}{r} 1\ 1\ 3 \\ \times\ \ \ 9 \\ \hline \end{array}$$

$$\begin{array}{r} 1\ 1\ 4 \\ \times\ \ \ 6 \\ \hline \end{array} \qquad \begin{array}{r} 1\ 1\ 4 \\ \times\ \ ♛ \\ \hline \end{array} \qquad \begin{array}{r} 1\ 1\ 4 \\ \times\ \ \ 7 \\ \hline \end{array} \qquad \begin{array}{r} 1\ 1\ 4 \\ \times\ \ \ 8 \\ \hline \end{array} \qquad \begin{array}{r} 1\ 1\ 4 \\ \times\ \ ♜ \\ \hline \end{array}$$

$$\begin{array}{r} 1\ 1\ 5 \\ \times\ \ ♜ \\ \hline \end{array} \qquad \begin{array}{r} 1\ 1\ 5 \\ \times\ \ \ 6 \\ \hline \end{array} \qquad \begin{array}{r} 1\ 1\ 5 \\ \times\ \ ♛ \\ \hline \end{array} \qquad \begin{array}{r} 1\ 1\ 5 \\ \times\ \ \ 8 \\ \hline \end{array} \qquad \begin{array}{r} 1\ 1\ 5 \\ \times\ \ \ 7 \\ \hline \end{array}$$

Ho Math Chess   何数棋谜   妈！我会棋谜式乘法啦！

Mom! I Learn Multiplication Using Math-Chess-Puzzles Connection!

Student's Name _____ Date _____

2007 - 2018 © Frank Ho, Amanda Ho, All rights reserved.   www.homathchess.com

ddd ✕ d with carrying

2 2 2	2 2 2	2 2 2	2 2 2	2 2 2
x 6	x 9	x 7	x 8	X ♖

2 2 3	2 2 3	2 2 3	2 2 3	2 2 3
x 6	X ♕	x 7	x 8	x 5

2 2 4	2 2 4	2 2 4	2 2 4	2 2 4
x 6	X ♖	x 7	x 8	x 5

2 2 5	2 2 5	2 2 5	2 2 5	2 2 5
X ♖	x 9	x 7	X ♕	x 5

2007 - 2018 © Frank Ho, Amanda Ho, All rights reserved.　　www.homathchess.com

ddd ✕ d with carrying

3 3 2	3 3 2	3 3 2	3 3 2	3 3 2
x 6	x ♕	x 7	x 8	x ♖

3 3 3	3 3 3	3 3 3	3 3 3	3 3 3
x 6	x ♕	x 7	x 8	x 5

3 3 4	3 3 4	3 3 4	3 3 4	3 3 4
x 6	x ♖	x 7	x 8	x 5

3 3 5	3 3 5	3 3 5	3 3 5	3 3 5
x 6	x 9	x ♖	x 8	x ♕

Ho Math Chess 何数棋谜 妈！我会棋谜式乘法啦！
Mom! I Learn Multiplication Using Math-Chess-Puzzles Connection!

Student's Name _____ Date _____

2007 - 2018 © Frank Ho, Amanda Ho, All rights reserved.   www.homathchess.com

ddd ✕ d with carrying

442	442	442	442	442
x 6	x 9	x 7	x 8	x 5

443	443	443	443	443
x 6	x ♕	x 7	x 8	x ♖

444	444	444	444	444
x 6	x ♕	x 7	x 8	x 5

445	445	445	445	445
x 6	x 8	x 7	x ♕	x ♖

2007 - 2018 © Frank Ho, Amanda Ho, All rights reserved.   www.homathchess.com

ddd ✕ d with carrying

$$
\begin{array}{r} 552 \\ \times\ \ 6 \\ \hline \end{array}
\qquad
\begin{array}{r} 552 \\ \times\ \ 8 \\ \hline \end{array}
\qquad
\begin{array}{r} 552 \\ \times\ \ 7 \\ \hline \end{array}
\qquad
\begin{array}{r} 552 \\ \times\ \ ♕ \\ \hline \end{array}
\qquad
\begin{array}{r} 552 \\ \times\ \ 5 \\ \hline \end{array}
$$

$$
\begin{array}{r} 553 \\ \times\ \ ♖ \\ \hline \end{array}
\qquad
\begin{array}{r} 553 \\ \times\ \ 8 \\ \hline \end{array}
\qquad
\begin{array}{r} 553 \\ \times\ \ 7 \\ \hline \end{array}
\qquad
\begin{array}{r} 553 \\ \times\ \ ♕ \\ \hline \end{array}
\qquad
\begin{array}{r} 553 \\ \times\ \ 6 \\ \hline \end{array}
$$

$$
\begin{array}{r} 554 \\ \times\ \ 6 \\ \hline \end{array}
\qquad
\begin{array}{r} 554 \\ \times\ \ 9 \\ \hline \end{array}
\qquad
\begin{array}{r} 554 \\ \times\ \ ♖ \\ \hline \end{array}
\qquad
\begin{array}{r} 554 \\ \times\ \ 8 \\ \hline \end{array}
\qquad
\begin{array}{r} 554 \\ \times\ \ 7 \\ \hline \end{array}
$$

$$
\begin{array}{r} 555 \\ \times\ \ 6 \\ \hline \end{array}
\qquad
\begin{array}{r} 555 \\ \times\ \ ♕ \\ \hline \end{array}
\qquad
\begin{array}{r} 555 \\ \times\ \ 7 \\ \hline \end{array}
\qquad
\begin{array}{r} 555 \\ \times\ \ 8 \\ \hline \end{array}
\qquad
\begin{array}{r} 555 \\ \times\ \ ♖ \\ \hline \end{array}
$$

2007 - 2018 © Frank Ho, Amanda Ho, All rights reserved.   www.homathchess.com

ddd ✕ d with carrying

```
  6 6 2        6 6 2        6 6 2        6 6 2        6 6 2
x     6      x     ♛      x     7      x     8      x     5


  6 6 3        6 6 3        6 6 3        6 6 3        6 6 3
x     6      x     8      x     7      x     ♛      x     ▣


  6 6 4        6 6 4        6 6 4        6 6 4        6 6 4
x     6      x     9      x     7      x     8      x     ▣


  6 6 5        6 6 5        6 6 5        6 6 5        6 6 5
x     6      x     ▣      x     7      x     8      x     ♛
```

331

Ho Math Chess   何数棋谜   妈!我会棋谜式乘法啦!
Mom! I Learn Multiplication Using Math-Chess-Puzzles Connection!

Student's Name _____ Date _____

2007 - 2018 © Frank Ho, Amanda Ho, All rights reserved.   www.homathchess.com

ddd ✕ d with carrying

7 7 2	7 7 2	7 7 2	7 7 2	7 7 2
x 6	x 9	x 7	x 8	x (rook)

7 7 3	7 7 3	7 7 3	7 7 3	7 7 3
x 6	x (queen)	x 7	x 8	x 5

7 7 4	7 7 4	7 7 4	7 7 4	7 7 4
x 6	x 9	x 7	x 8	x (rook)

7 7 5	7 7 5	7 7 5	7 7 5	7 7 5
x 6	x (queen)	x 7	x 8	x 5

Ho Math Chess　何数棋谜　妈!我会棋谜式乘法啦!
Mom! I Learn Multiplication Using Math-Chess-Puzzles Connection!

Student's Name _____ Date _____

2007 - 2018 © Frank Ho, Amanda Ho, All rights reserved.　　www.homathchess.com

ddd ✕ d with carrying

882	882	882	882	882
x　6	x　♜	x　7	x　8	x　♛

883	883	883	883	883
x　6	x　♛	x　7	x　♜	x　8

884	884	884	884	884
x　♜	x　9	x　7	x　8	x　6

885	885	885	885	885
x　6	x　♛	x　7	x　8	x　♜

2007 - 2018 © Frank Ho, Amanda Ho, All rights reserved.   www.homathchess.com

ddd ✕ d with carrying

992	992	999	992	992
x 6	x ♛	x ♜	x 8	x 7

993	993	993	993	993
x 6	x ♜	x 7	x 8	x ♛

994	994	994	994	994
x 6	x 8	x 7	x ♛	x ♜

995	995	995	995	995
x 6	x ♛	x ♜	x 8	x 7

2007 - 2018 © Frank Ho, Amanda Ho, All rights reserved.　　www.homathchess.com

Changing the order of d X dd to dd X d

$$
\begin{array}{r}
3 \\
\times\ 15 \\
\hline
\end{array}
\qquad
\begin{array}{r}
15 \\
\times\ \text{♗} \\
\hline
\end{array}
\qquad
\begin{array}{r}
3 \\
\times\ 19 \\
\hline
\end{array}
\qquad
\begin{array}{r}
19 \\
\times\ \text{♗} \\
\hline
\end{array}
$$

Comparison of 3 X 1234 is changed to the offer of 1234 X 3.

Regular way as presented	Bigger number X smaller number
X　　1　2　3　4 ♗ □□□□ □□□ □□ □ ――――― □□□□	1　2　3　4 X　　　♗ □□□□

2007 - 2018 © Frank Ho, Amanda Ho, All rights reserved.   www.homathchess.com

dd X d0s (X by multiples of 10's, equivalent to ÷ multiples of 0.1)

Place the number ending with 0s as the second factor and just bring down 0s.

$$\begin{array}{r} 2\ 1 \\ \times\ 2\ 0 \\ \hline \square\square \\ \square\square \\ \hline \square\square\square \end{array}$$

$$\begin{array}{r} 2\ 1 \\ \times\ \ 2\ 0 \\ \hline \square\square\ 0 \end{array}$$

Just bring down one zero.

$$\begin{array}{r} 1\ 9 \\ \times\ 2\ 0\ 0 \\ \hline \square\square \\ \square\square \\ \square\square \\ \hline \square\square\square\square \end{array}$$

$$\begin{array}{r} 1\ 9 \\ \times\ \ 2\ 0\ 0 \\ \hline \square\square 00 \end{array}$$

Just bring down two zeros.

2 X 1000 = _____   999 X 1000 = _____

300 X 1000 = _____   909 X 1000 = _____

2100 X 100 = _____   101 X 1000 = _____

99 X 1000 = _____   123 X 1000 = _____

10000 X 540 = _____   10000 X 111 = _____

2007 - 2018 © Frank Ho, Amanda Ho, All rights reserved.   www.homathchess.com

dd X d0s (X by multiples of 10's, equivalent to ÷ multiples of 0.1)

Place the number ending with 0s as the second factor and just bring down 0s.

300 X 123 = 123 X 300?

Regular way	Bring down zeros.
3 0 0 X 1 2 3 ☐☐☐ ☐☐☐ ☐☐☐ ――――― ☐☐☐☐☐	1 2 3 X 3 0 0 ☐☐☐☐☐

2007 - 2018 © Frank Ho, Amanda Ho, All rights reserved.   www.homathchess.com

dd X d0s (X by multiples of 10's, equivalent to ÷ multiples of 0.1)

Place the number ending with 0s as the second factor and just bring down 0s.

123 X 12000	$\begin{array}{r} 1\ 2\ 3 \\ \underline{X\ \ 1\ 2\ 0\ 0\ 0} \end{array}$
13000 X 14	
234 X 12000	

dd $\times$ d0

28 x 20	28 x 20	23 x 50	23 x 50
46 x 20	39 x 60	79 x 50	17 x 40
26 x 80	53 x 30	47 x 60	78 x 70

dd × d0

$$\begin{array}{r} 43 \\ \times\ 30 \\ \hline \end{array} \qquad \begin{array}{r} 53 \\ \times\ 40 \\ \hline \end{array} \qquad \begin{array}{r} 63 \\ \times\ 30 \\ \hline \end{array} \qquad \begin{array}{r} 27 \\ \times\ 90 \\ \hline \end{array}$$

$$\begin{array}{r} 76 \\ \times\ 70 \\ \hline \end{array} \qquad \begin{array}{r} 58 \\ \times\ 60 \\ \hline \end{array} \qquad \begin{array}{r} 45 \\ \times\ 50 \\ \hline \end{array} \qquad \begin{array}{r} 53 \\ \times\ 80 \\ \hline \end{array}$$

$$\begin{array}{r} 87 \\ \times\ 90 \\ \hline \end{array} \qquad \begin{array}{r} 39 \\ \times\ 80 \\ \hline \end{array} \qquad \begin{array}{r} 95 \\ \times\ 50 \\ \hline \end{array} \qquad \begin{array}{r} 67 \\ \times\ 40 \\ \hline \end{array}$$

dd × d0

12 X 20 =   32 X 30 =   52 X 50 =

33 X 40 =   52 X 60 =   63 X 20 =

32 X 70 =   54 X 40 =   17 X 80 =

27 X 90 =   62 X 60 =   72 X 30 =

2007 - 2018 © Frank Ho, Amanda Ho, All rights reserved.   www.homathchess.com

dd ✕ d0s

$$
\begin{array}{r} 27 \\ \times\ 400 \\ \hline \end{array}
\qquad
\begin{array}{r} 29 \\ \times\ 500 \\ \hline \end{array}
\qquad
\begin{array}{r} 75 \\ \times\ 700 \\ \hline \end{array}
\qquad
\begin{array}{r} 96 \\ \times\ 400 \\ \hline \end{array}
$$

$$
\begin{array}{r} 19 \\ \times\ 300 \\ \hline \end{array}
\qquad
\begin{array}{r} 73 \\ \times\ 400 \\ \hline \end{array}
\qquad
\begin{array}{r} 49 \\ \times\ 600 \\ \hline \end{array}
\qquad
\begin{array}{r} 35 \\ \times\ 900 \\ \hline \end{array}
$$

$$
\begin{array}{r} 91 \\ \times\ 800 \\ \hline \end{array}
\qquad
\begin{array}{r} 53 \\ \times\ 700 \\ \hline \end{array}
\qquad
\begin{array}{r} 76 \\ \times\ 500 \\ \hline \end{array}
\qquad
\begin{array}{r} 83 \\ \times\ 900 \\ \hline \end{array}
$$

2007 - 2018 © Frank Ho, Amanda Ho, All rights reserved.   www.homathchess.com

dd × d0s

42 X 600 =   37 X 200 =   98 X 400 =

28 X 700 =   48 X 300 =   43 X 400 =

17 X 900 =   27 X 500 =   61 X 200 =

24 X 900 =   73 X 500 =   34 X 400 =

2007 - 2018 © Frank Ho, Amanda Ho, All rights reserved.   www.homathchess.com

dd × dd multiplication concepts

Horizontal multiplication

$23 \times 24 = 23 \times (4 + 20) = 23 \times 4 + 23 \times 20 = 92 + 460 = 552$

Vertical multiplication

It is extremely important for teacher to explain the concepts of why and how multiplication is done. Repeated drills without explanations only add prolong learning curve which could have been reduced if concepts such as the following had been explained.

1. Why line up ones multiplication at the rightmost position? Because it is ones place.
2. Why a 0 is placed at the ones place when doing tens place multiplication? Because tens place multiplication always has a 0 at ones place such as 20 in the example, the ones place value is 0.
3. How is the horizontal multiplication related to the vertical multiplication? The vertical multiplication is using the concept of distributive law to do the work but written in a vertical way. The following is an example.

$$
\begin{array}{r}
27 \\
\times\ 36 \\
\hline
42 \\
120 \\
21 \\
+\ 600 \\
\hline
972 \\
\end{array}
$$

2007 - 2018 © Frank Ho, Amanda Ho, All rights reserved.   www.homathchess.com

Step 1	Step 2
$\begin{array}{r} 2\ 3 \\ \times\ \ \ 4 \\ \hline 9\ 2 \end{array}$	$\begin{array}{r} 2\ 3 \\ \times\ \ 2\ 0 \\ \hline 4\ 6\ 0 \end{array}$

Step 3: The answer is 92 + 460 = 552

$25 \times 24 = 25 \times (4 + 20) = 25 \times 4 + 25 \times 20 =$

$24 \times 25 = 24 \times (5 + 20) = 24 \times 5 + 24 \times 20 =$

$25 \times 36 = 25 \times (6 + 30) = 25 \times 6 + 25 \times 30 =$

$36 \times 25 = 36 \times (5 + 20) = 36 \times 5 + 36 \times 20 =$

$27 \times 28 = 27 \times (8 + 20) = 27 \times 8 + 27 \times 20 =$

$28 \times 27 = 28 \times (7 + 20) = 28 \times 7 + 28 \times 20 =$

2007 - 2018 © Frank Ho, Amanda Ho, All rights reserved.　www.homathchess.com

dd ✕ dd with carrying

```
    1                          1                    2
   12        21              23                24
 x 16      x 16            x 16              x 16
 ─────     ─────           ─────             ─────
   72      □□□             □□□               □□□
   12      □□              □□                □□
 ─────     ─────           ─────             ─────
  192      □□□             □□□               □□□
```

```
   15        16              17                18
 x 15      x 27            x 28              x 29
 ─────     ─────           ─────             ─────
  □□       □□□             □□□               □□□
  □□       □□              □□                □□
 ─────     ─────           ─────             ─────
  □□□      □□□             □□□               □□□
```

dd × dd with carrying

```
    1 5          1 5          1 5              1 5
  x 1 6        x 1 7        x 1 8           x 1 9
  [ ][ ]      [ ][ ][ ]    [ ][ ][ ]       [ ][ ][ ]
  [ ][ ]      [ ][ ]        [ ][ ]          [ ][ ]
  ―――――       ―――――        ―――――          ―――――
  [ ][ ][ ]   [ ][ ][ ]    [ ][ ][ ]       [ ][ ][ ]
```

```
    1 5          1 8          1 9              2 1
  x 1 6        x 1 7        x 1 8           x 1 9
  [ ][ ]      [ ][ ][ ]    [ ][ ][ ]       [ ][ ][ ]
  [ ][ ]      [ ][ ]        [ ][ ]          [ ][ ]
  ―――――       ―――――        ―――――          ―――――
  [ ][ ][ ]   [ ][ ][ ]    [ ][ ][ ]       [ ][ ][ ]
```

2007 - 2018 © Frank Ho, Amanda Ho, All rights reserved.   www.homathchess.com

dd × dd with carrying

```
    2 5        2 6        2 7        2 8
  x 1 6      x  1 7     x  1 9     x  2 1
  □□□        □□□        □□□        □□□
   □□         □□         □□         □□
  ─────      ─────      ─────      ─────
  □□□        □□□        □□□        □□□
```

```
    2 6        2 7        2 8        2 9
  x 1 6      x  2 7     x  2 8     x  2 9
  □□□        □□□        □□□        □□□
   □□         □□         □□         □□
  ─────      ─────      ─────      ─────
  □□□        □□□        □□□        □□□
```

Ho Math Chess　何数棋谜　妈！我会棋谜式乘法啦！
Mom! I Learn Multiplication Using Math-Chess-Puzzles Connection!

Student's Name _____ Date _____

2007 - 2018 © Frank Ho, Amanda Ho, All rights reserved.　　www.homathchess.com

dd × dd with carrying

```
    37          38          39              49
  x 16        x 17        x 19          x  21
  □□□         □□□         □□□            □□
   □□          □□          □□            □□
  □□□         □□□         □□□           □□□□
```

```
    48          49          51              51
  x 16        x 17        x 18          x  19
  □□□         □□□         □□□           □□□
   □□          □□          □□            □□
  □□□         □□□         □□□           □□□
```

Ho Math Chess　　何数棋谜　妈！我会棋谜式乘法啦！
Mom! I Learn Multiplication Using Math-Chess-Puzzles Connection!
Student's Name _____ Date _____
2007 - 2018 © Frank Ho, Amanda Ho, All rights reserved.　　www.homathchess.com

dd × dd with carrying

$$
\begin{array}{r}
2\ 9 \\
\times\ 1\ 6 \\
\hline
\end{array}
\qquad
\begin{array}{r}
2\ 9 \\
\times\ 1\ 7 \\
\hline
\end{array}
\qquad
\begin{array}{r}
2\ 9 \\
\times\ 1\ 8 \\
\hline
\end{array}
\qquad
\begin{array}{r}
2\ 9 \\
\times\ 1\ 9 \\
\hline
\end{array}
$$

$$
\begin{array}{r}
3\ 5 \\
\times\ 1\ 6 \\
\hline
\end{array}
\qquad
\begin{array}{r}
3\ 5 \\
\times\ 1\ 7 \\
\hline
\end{array}
\qquad
\begin{array}{r}
3\ 5 \\
\times\ 1\ 8 \\
\hline
\end{array}
\qquad
\begin{array}{r}
3\ 5 \\
\times\ 1\ 9 \\
\hline
\end{array}
$$

Ho Math Chess　何数棋谜　妈！我会棋谜式乘法啦！
Mom! I Learn Multiplication Using Math-Chess-Puzzles Connection!

Student's Name _____ Date _____

2007 - 2018 © Frank Ho, Amanda Ho, All rights reserved.　www.homathchess.com

dd × dd with carrying

$$
\begin{array}{r}
36 \\
\times\ 16 \\
\hline
\end{array}
\qquad
\begin{array}{r}
36 \\
\times\ 17 \\
\hline
\end{array}
\qquad
\begin{array}{r}
36 \\
\times\ 18 \\
\hline
\end{array}
\qquad
\begin{array}{r}
36 \\
\times\ 19 \\
\hline
\end{array}
$$

$$
\begin{array}{r}
37 \\
\times\ 16 \\
\hline
\end{array}
\qquad
\begin{array}{r}
37 \\
\times\ 17 \\
\hline
\end{array}
\qquad
\begin{array}{r}
37 \\
\times\ 18 \\
\hline
\end{array}
\qquad
\begin{array}{r}
37 \\
\times\ 19 \\
\hline
\end{array}
$$

2007 - 2018 © Frank Ho, Amanda Ho, All rights reserved.   www.homathchess.com

dd × dd with carrying

```
    3 8        3 9            4 6          5 6
x   2 6    x   2 7        x   2 8      x   2 9
  □ □ □      □ □ □          □ □ □        □ □ □
    □ □        □ □            □ □        □ □ □
  □ □ □      □ □ □ □        □ □ □ □      □ □ □ □
```

```
    5 7        5 8          5 9          6 7
x   2 6    x   2 7      x   2 8      x   2 9
  □ □ □      □ □ □        □ □ □        □ □ □
  □ □ □      □ □ □        □ □ □        □ □ □
  □ □ □ □    □ □ □ □      □ □ □ □      □ □ □ □
```

dd × dd with carrying

```
    1             2             1             2
   1 2          2 1          2 3          2 4
 x 1 6        x 1 6        x 1 6        x 1 6
 ───────      ───────      ───────      ───────
   7 2        □□□          □□□          □□□
   1 2        □□           □□           □□
 ───────      ───────      ───────      ───────
 1 9 2        □□□          □□□          □□□
```

```
   1 5          1 6          1 7          1 8
 x 1 5        x 2 7        x 2 8        x 2 9
 ───────      ───────      ───────      ───────
   □□         □□□          □□□          □□□
   □□         □□           □□           □□
 ───────      ───────      ───────      ───────
   □□□        □□□          □□□          □□□
```

2007 - 2018 © Frank Ho, Amanda Ho, All rights reserved.　　www.homathchess.com

dd × dd with carrying

```
    1 5        1 5        1 5        1 5
  x 1 6      x   1 7    x   1 8    x   1 9
  [ ][ ]     [ ][ ][ ]  [ ][ ][ ]  [ ][ ][ ]
  [ ][ ]     [ ][ ]     [ ][ ]     [ ][ ]
  ─────      ─────      ─────      ─────
  [ ][ ][ ]  [ ][ ][ ]  [ ][ ][ ]  [ ][ ][ ]
```

```
    1 5        1 8        1 9        2 1
  x 1 6      x   1 7    x   1 8    x   1 9
  [ ][ ]     [ ][ ][ ]  [ ][ ][ ]  [ ][ ][ ]
  [ ][ ]     [ ][ ]     [ ][ ]     [ ][ ]
  ─────      ─────      ─────      ─────
  [ ][ ][ ]  [ ][ ][ ]  [ ][ ][ ]  [ ][ ][ ]
```

Ho Math Chess   何数棋谜 妈！我会棋谜式乘法啦！
Mom! I Learn Multiplication Using Math-Chess-Puzzles Connection!

Student's Name _____ Date _____

2007 - 2018 © Frank Ho, Amanda Ho, All rights reserved.   www.homathchess.com

dd ✕ dd without carrying

$$
\begin{array}{r} 47 \\ \times\,1\,1 \\ \hline \end{array}
\qquad
\begin{array}{r} 32 \\ \times\,2\,2 \\ \hline \end{array}
\qquad
\begin{array}{r} 23 \\ \times\,3\,3 \\ \hline \end{array}
\qquad
\begin{array}{r} 38 \\ \times\,1\,1 \\ \hline \end{array}
$$

$$
\begin{array}{r} 42 \\ \times\,2\,2 \\ \hline \end{array}
\qquad
\begin{array}{r} 32 \\ \times\,2\,2 \\ \hline \end{array}
\qquad
\begin{array}{r} 29 \\ \times\,1\,1 \\ \hline \end{array}
\qquad
\begin{array}{r} 37 \\ \times\,1\,1 \\ \hline \end{array}
$$

$$
\begin{array}{r} 27 \\ \times\,1\,1 \\ \hline \end{array}
\qquad
\begin{array}{r} 33 \\ \times\,2\,2 \\ \hline \end{array}
\qquad
\begin{array}{r} 37 \\ \times\,1\,1 \\ \hline \end{array}
\qquad
\begin{array}{r} 14 \\ \times\,2\,2 \\ \hline \end{array}
$$

2007 - 2018 © Frank Ho, Amanda Ho, All rights reserved.　www.homathchess.com

dd ✕ dd with carrying

$$
\begin{array}{r} 45 \\ \times\,25 \\ \hline \end{array}
\qquad
\begin{array}{r} 36 \\ \times\,46 \\ \hline \end{array}
\qquad
\begin{array}{r} 27 \\ \times\,57 \\ \hline \end{array}
\qquad
\begin{array}{r} 38 \\ \times\,48 \\ \hline \end{array}
$$

□□□
□□□

$$
\begin{array}{r} 33 \\ \times\,22 \\ \hline \end{array}
\qquad
\begin{array}{r} 22 \\ \times\,33 \\ \hline \end{array}
\qquad
\begin{array}{r} 24 \\ \times\,22 \\ \hline \end{array}
\qquad
\begin{array}{r} 44 \\ \times\,22 \\ \hline \end{array}
$$

2007 - 2018 © Frank Ho, Amanda Ho, All rights reserved.   www.homathchess.com

dd × dd with carrying

```
   2 7        3 6        3 7        1 8
 x 2 7      x 5 6      x 6 7      x 7 8
 □□□        □□□        □□□        □□□
 □□□        □□□        □□□        □□□
```

```
   5 9        2 6        2 7        8 8
 x 2 9      x 4 6      x 3 7      x 6 8
 □□□        □□□        □□□        □□□
 □□□        □□□        □□□        □□□
```

```
   1 7        2 6        2 5        4 8
 x 2 1      x 4 2      x 5 6      x 3 6
```

Ho Math Chess 何数棋谜 妈！我会棋谜式乘法啦！
Mom! I Learn Multiplication Using Math-Chess-Puzzles Connection!

Student's Name _____ Date _____

2007 - 2018 © Frank Ho, Amanda Ho, All rights reserved.   www.homathchess.com

dd ✕ dd with carrying

```
   4 5        1 2        2 9        3 7
 x 1 9      x 4 7      x 5 5      x 2 3
```

```
   2 7        1 1        5 7        1 4
 x 1 1      x 5 6      x 2 6      x 5 2
```

```
   3 2        2 3        2 4        2 4
 x 2 1      x 4 2      x 3 2      x 1 3
```

Ho Math Chess　何数棋谜　妈！我会棋谜式乘法啦！
Mom! I Learn Multiplication Using Math-Chess-Puzzles Connection!
Student's Name _____ Date _____
2007 - 2018 © Frank Ho, Amanda Ho, All rights reserved.　www.homathchess.com

dd ✕ dd with carrying

```
    47          36          27          38
  x 28        x 46        x 56        x 48
  □□□         □□□         □□□         □□□
  □□□         □□□         □□□         □□□
```

```
    45          32          29          37
  x 29        x 47        x 55        x 43
  □□□         □□□         □□□         □□□
  □□□         □□□         □□□         □□□
```

```
    27          33          37          18
  x 29        x 56        x 66        x 72
  □□□         □□□         □□□         □□□
  □□□         □□□         □□□         □□□
```

2007 - 2018 © Frank Ho, Amanda Ho, All rights reserved.　　www.homathchess.com

dd ✕ dd with carrying

```
   5 7          2 6          2 7          8 8
 x 2 9        x 4 6        x 3 6        x 6 6
 □□□          □□□          □□□          □□□
 □□□          □□□          □□□          □□□
```

```
   5 1          3 6          2 9          6 8
 x 2 8        x 4 2        x 3 1        x 6 2
```

```
   6 3          7 8          8 4          7 4
 x 8 4        x 4 6        x 2 5        x 6 8
```

2007 - 2018 © Frank Ho, Amanda Ho, All rights reserved.　　www.homathchess.com

dd ✕ dd with carrying

```
    4 7          3 6          2 7          3 8
  x 2 8        x 4 6        x 5 6        x 4 6
  □□□          □□□          □□□          □□□
  □□□          □□□          □□□          □□□
```

```
    4 5          3 2          2 9          3 7
  x 2 9        x 4 7        x 5 5        x 4 3
```

```
    2 7          3 3          3 7          1 8
  x 2 9        x 5 6        x 6 6        x 7 2
```

2007 - 2018 © Frank Ho, Amanda Ho, All rights reserved.　www.homathchess.com

dd ✕ dd with carrying

12 X 28 =   32 X 27 =   52 X 35 =

33 X 45 =   52 X 25 =   63 X 21 =

32 X 27 =   54 X 18 =   17 X 48 =

27 X 83 =   62 X 24 =   72 X 32 =

2007 - 2018 © Frank Ho, Amanda Ho, All rights reserved.   www.homathchess.com

dd ✕ dd with carrying

42 X 38 =   35 X 25 =   52 X 25 =

38 X 45 =   48 X 25 =   69 X 21 =

76 X 27 =   36 X 41 =   95 X 28 =

62 X 73 =   53 X 26 =   74 X 36 =

2007 - 2018 © Frank Ho, Amanda Ho, All rights reserved.　www.homathchess.com

dd ✕ dd with carrying

54 X 23 =   62 X 23 =   71 X 85 =

53 X 45 =   82 X 75 =   63 X 48 =

81 X 37 =   76 X 52 =   49 X 68 =

77 X 84 =   93 X 54 =   82 X 39 =

Ho Math Chess　何数棋谜　妈！我会棋谜式乘法啦！
Mom! I Learn Multiplication Using Math-Chess-Puzzles Connection!

Student's Name _____ Date _____

2007 - 2018 © Frank Ho, Amanda Ho, All rights reserved.　www.homathchess.com

dd $\times$ 10s

$35 \times 10 =$　　　　　$57 \times 10 =$　　　　　$26 \times 10 =$

$53 \times 100 =$　　　　$21 \times 100 =$　　　　$64 \times 100 =$

$84 \times 1000 =$　　　$86 \times 1000 =$　　　$24 \times 1000 =$

$64 \times 10 =$　　　　　$57 \times 100 =$　　　　$97 \times 1000 =$

Ho Math Chess　　何数棋谜　妈!我会棋谜式乘法啦!
Mom! I Learn Multiplication Using Math-Chess-Puzzles Connection!
Student's Name _____ Date _____

2007 - 2018 © Frank Ho, Amanda Ho, All rights reserved.　　www.homathchess.com

dd $\times$ 5

For advanced students only.

42 X 5

=21 X 2 X 5

=21 X 10

=210

38 X 5 =

52 X 5 =

36 X 5 =

48 X 5 =

64 X 5 =

76 X 5 =

56 X 5 =

96 X 5 =

62 X 5 =

54 X 5 =

74 X 5 =

2007 - 2018 © Frank Ho, Amanda Ho, All rights reserved.   www.homathchess.com

dd $\times$ 25

For advanced students only.

8 X 25

=2 X 4 X 25

=2 X 100

=200

36 X 25 =

48 X 25 =

24 X 25 =

20 X 25 =

16 X 25 =

12 X 25 =

32 X 25 =

28 X 25 =

44 X 25 =

40 X 25 =

56 X 25 =

2007 - 2018 © Frank Ho, Amanda Ho, All rights reserved.   www.homathchess.com

dd $\times$ 11

23 X 11

=23 X (10+1)

=23 X 10+23 X 1

=230+23

=253 (The other way is to use 2 2+3 3)

32 X 11 =

52 X 11 =

31 X 11 =

45 X 11 =

44 X 11 =

71 X 11 =

56 X 11 =

96 X 11 =

68 X 11 =

57 X 11 =

74 X 11 =

Mom! I Learn Multiplication Using Math-Chess-Puzzles Connection!

Student's Name _____ Date _____

2007 - 2018 © Frank Ho, Amanda Ho, All rights reserved.   www.homathchess.com

ddd ✕ dd with carrying

$$
\begin{array}{r}
125 \\
\times\ 16 \\
\hline
\square\square\square \\
\square\square \\
\hline
\square\square\square\square \\
\end{array}
\qquad
\begin{array}{r}
126 \\
\times\ 17 \\
\hline
\square\square\square \\
\square\square \\
\hline
\square\square\square\square \\
\end{array}
\qquad
\begin{array}{r}
127 \\
\times\ 15 \\
\hline
\square\square\square \\
\square\square \\
\hline
\square\square\square\square \\
\end{array}
$$

$$
\begin{array}{r}
234 \\
\times\ 56 \\
\hline
\square\square\square\square \\
\square\square\square\square \\
\hline
\square\square\square\square\square \\
\end{array}
\qquad
\begin{array}{r}
235 \\
\times\ 57 \\
\hline
\square\square\square\square \\
\square\square\square\square \\
\hline
\square\square\square\square\square \\
\end{array}
\qquad
\begin{array}{r}
236 \\
\times\ 58 \\
\hline
\square\square\square\square \\
\square\square\square\square \\
\hline
\square\square\square\square\square \\
\end{array}
$$

Ho Math Chess 何数棋谜 妈!我会棋谜式乘法啦!
Mom! I Learn Multiplication Using Math-Chess-Puzzles Connection!

Student's Name _____ Date _____

2007 - 2018 © Frank Ho, Amanda Ho, All rights reserved.   www.homathchess.com

ddd ✕ dd with carrying

$$345 \times 65$$

$$345 \times 67$$

$$345 \times 68$$

$$666 \times 66$$

$$777 \times 77$$

$$888 \times 88$$

2007 - 2018 © Frank Ho, Amanda Ho, All rights reserved.　　www.homathchess.com

ddd $\times$ dd with carrying

$$
\begin{array}{r}
345 \\
\times\ \ 77 \\
\hline
\end{array}
\qquad
\begin{array}{r}
345 \\
\times\ \ 88 \\
\hline
\end{array}
\qquad
\begin{array}{r}
345 \\
\times\ \ 99 \\
\hline
\end{array}
$$

$$
\begin{array}{r}
999 \\
\times\ \ 77 \\
\hline
\end{array}
\qquad
\begin{array}{r}
999 \\
\times\ \ 88 \\
\hline
\end{array}
\qquad
\begin{array}{r}
999 \\
\times\ \ 99 \\
\hline
\end{array}
$$

ddd ✕ dd with carrying

```
   567          765          675
 x  77        x  88        x  99
 □□□□         □□□□         □□□□
 □□□□         □□□□         □□□□
 □□□□□        □□□□□        □□□□□

   678          876          786
 x  77        x  88        x  99
 □□□□         □□□□         □□□□
 □□□□         □□□□         □□□□
 □□□□□        □□□□□        □□□□□
```

2007 - 2018 © Frank Ho, Amanda Ho, All rights reserved.　　www.homathchess.com

ddd ✕ dd without carrying

```
   472        425        432        384
 x 11      x 22      x 33      x 11


   323        421        224        352
 x 22      x 22      x 11      x 11


   286        412        374        243
 x 11      x 22      x 11      x 22
```

2007 - 2018 © Frank Ho, Amanda Ho, All rights reserved.   www.homathchess.com

ddd ✕ dd with carrying

$$
\begin{array}{r} 474 \\ \times\ 15 \\ \hline \end{array}
\qquad
\begin{array}{r} 316 \\ \times\ 36 \\ \hline \end{array}
\qquad
\begin{array}{r} 237 \\ \times\ 28 \\ \hline \end{array}
\qquad
\begin{array}{r} 589 \\ \times\ 31 \\ \hline \end{array}
$$

$$
\begin{array}{r} 457 \\ \times\ 52 \\ \hline \end{array}
\qquad
\begin{array}{r} 628 \\ \times\ 27 \\ \hline \end{array}
\qquad
\begin{array}{r} 275 \\ \times\ 61 \\ \hline \end{array}
\qquad
\begin{array}{r} 874 \\ \times\ 39 \\ \hline \end{array}
$$

$$
\begin{array}{r} 577 \\ \times\ 61 \\ \hline \end{array}
\qquad
\begin{array}{r} 368 \\ \times\ 42 \\ \hline \end{array}
\qquad
\begin{array}{r} 779 \\ \times\ 43 \\ \hline \end{array}
\qquad
\begin{array}{r} 147 \\ \times\ 29 \\ \hline \end{array}
$$

Mom! I Learn Multiplication Using Math-Chess-Puzzles Connection!

Student's Name _____ Date _____

2007 - 2018 © Frank Ho, Amanda Ho, All rights reserved.   www.homathchess.com

ddd ✕ dd with carrying

159 X 64 =   357 X 24 =   684 X 86 =

268 X 57 =   842 X 17 =   368 X 43 =

251 X 74 =   961 X 19 =   254 X 24 =

741 X 27 =   217 X 14 =   364 X 43 =

2007 - 2018 © Frank Ho, Amanda Ho, All rights reserved.   www.homathchess.com

ddd × dd with carrying

148 X 24 =   572 X 47 =   351 X 34 =

381 X 56 =   729 X 72 =   349 X 764=

941 X 48 =   149 X 24 =   324 X 27 =

327 X 34 =   243 X 63 =   751 X 82 =

2007 - 2018 © Frank Ho, Amanda Ho, All rights reserved.　www.homathchess.com

d0d $\times$ dd

305 X 47 =   507 X 55 =   805 X 51 =

609 X 34 =   709 X 78 =   501 X 39 =

207 X 49 =   608 X 56 =   901 X 27 =

206 X 91 =   604 X 28 =   908 X 62 =

2007 - 2018 © Frank Ho, Amanda Ho, All rights reserved. www.homathchess.com

dd X d0d

The 0s in the middle of a factor do not change the sum, so no product needs to be done.

Long form

```
  1 2 3
x   1 0 1
```

Short form

```
  1 2 3
x   1 0 1
```
(Do not multiply 0)

```
  2 1        2 1         1 9          1 9
x 2 0 1    x 3 0 1     x 4 0 1      x 5 0 1
```

Mom! I Learn Multiplication Using Math-Chess-Puzzles Connection!

Student's Name _____ Date _____

2007 - 2018 © Frank Ho, Amanda Ho, All rights reserved.   www.homathchess.com

d0d × dd

$$\begin{array}{r} 6\,0\,3 \\ \times\ 2\,3 \\ \hline 1\,8\,0\,9 \\ \underline{1\,2\,0\,6} \\ 1\,3\,8\,6\,9 \end{array}$$

$$\begin{array}{r} 7\,0\,8 \\ \times\ 3\,7 \\ \hline \end{array}$$

$$\begin{array}{r} 5\,0\,3 \\ \times\ 4\,8 \\ \hline \end{array}$$

$$\begin{array}{r} 8\,0\,5 \\ \times\ 5\,1 \\ \hline \end{array}$$

$$\begin{array}{r} 7\,0\,4 \\ \times\ 7\,2 \\ \hline \end{array}$$

$$\begin{array}{r} 2\,0\,9 \\ \times\ 8\,7 \\ \hline \end{array}$$

$$\begin{array}{r} 3\,0\,7 \\ \times\ 6\,7 \\ \hline \end{array}$$

$$\begin{array}{r} 4\,0\,6 \\ \times\ 3\,5 \\ \hline \end{array}$$

$$\begin{array}{r} 5\,0\,4 \\ \times\ 5\,1 \\ \hline \end{array}$$

$$\begin{array}{r} 6\,0\,4 \\ \times\ 7\,2 \\ \hline \end{array}$$

$$\begin{array}{r} 2\,0\,9 \\ \times\ 9\,3 \\ \hline \end{array}$$

$$\begin{array}{r} 8\,0\,7 \\ \times\ 4\,9 \\ \hline \end{array}$$

d0d × dd

502	407	705	903
x 37	x 42	x 51	x 45

205	706	408	105
x 17	x 17	x 38	x 18

802	708	204	809
x 43	x 71	x 86	x 16

2007 - 2018 © Frank Ho, Amanda Ho, All rights reserved.   www.homathchess.com

d0d ✕ dd

609 X 47 =   304 X 24 =   605 X 38 =

508 X 57 =   702 X 45 =   308 X 29 =

801 X 43 =   501 X 24 =   804 X 37 =

401 X 52 =   807 X 41 =   204 X 62 =

2007 - 2018 © Frank Ho, Amanda Ho, All rights reserved.   www.homathchess.com

dd X d0d

The 0s in the middle of a factor do not change the sum, so no product needs to be done.

```
    2 0 1        2 0 1        1 0 9        1 0 9
 x    2 0 1    x   3 0 1    x   4 0 1    x   5 0 1
      □ □ □        □ □ □        □ □ □        □ □ □
    □ □ □ □      □ □ □ □      □ □ □ □      □ □ □ □
    □ □ □ □ □    □ □ □ □ □    □ □ □ □ □    □ □ □ □ □
```

```
    2 0 0 1      2 0 0 1      3 0 0 1      4 0 0 1
 x   2 0 0 1   x  3 0 0 1   x  4 0 0 1   x 5 0 0 1
```

Mom! I Learn Multiplication Using Math-Chess-Puzzles Connection!

Student's Name _____ Date _____

2007 - 2018 © Frank Ho, Amanda Ho, All rights reserved.　　www.homathchess.com

1s X 1s

$11 \times 11 =$　　　　$111 \times 11 =$　　　　$1111 \times 11 =$

$11111 \times 11 =$　　　$111111 \times 11 =$　　　$1111111 \times 11 =$

$111 \times 111 =$　　　$1111 \times 1111 =$　　　$11111 \times 11111 =$

$222 \times 111 =$　　　$333 \times 111 =$　　　$444 \times 111 =$

2007 - 2018 © Frank Ho, Amanda Ho, All rights reserved.   www.homathchess.com

1s X 1s

101 X 5 =   101 X 8 =   101 X 6 =

101 X 11 =   101 X 26 =   101 X 75 =

101 X 111 =   101 X 123 =   101 X 532 =

101 X 45 =   101 X 78 =   101 X 642 =

2007 - 2018 © Frank Ho, Amanda Ho, All rights reserved.   www.homathchess.com

5s X 5s

15 X 15 =   25 X 25 =   35 X 35 =

45 X 45 =   55 X 55 =   65 X 65 =

75 X 75 =   85 X 85 =   95 X 95 =

25 X 25 =   45 X 45 =   75 X 75 =

Ho Math Chess   何数棋谜   妈！我会棋谜式乘法啦！
Mom! I Learn Multiplication Using Math-Chess-Puzzles Connection!
Student's Name _____ Date _____
2007 - 2018 © Frank Ho, Amanda Ho, All rights reserved.   www.homathchess.com

d $\times$ d $\times$ d

$8 \times 6 \times 9$
$= 48 \times 9$
$= 432$

$11 \times 25 \times 9$
$=$

$13 \times 14 \times 8$
$=$

$21 \times 15 \times 34$
$=$

$32 \times 11 \times 72$
$=$

$17 \times 18 \times 35$
$=$

$24 \times 32 \times 17$
$=$

$27 \times 31 \times 52$
$=$

$33 \times 47 \times 41$
$=$

$52 \times 13 \times 26$
$=$

$43 \times 82 \times 24$
$=$

$48 \times 41 \times 24$
$=$

2007 - 2018 © Frank Ho, Amanda Ho, All rights reserved.　　www.homathchess.com

Replace the ? with a number.

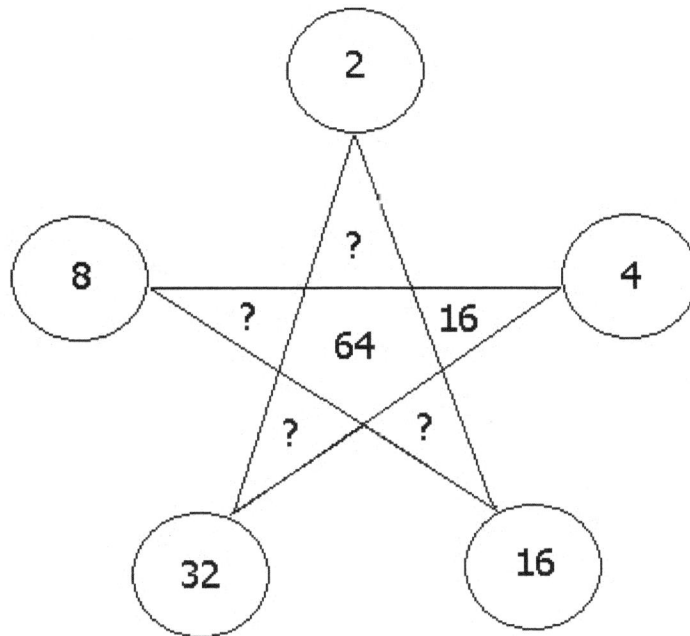

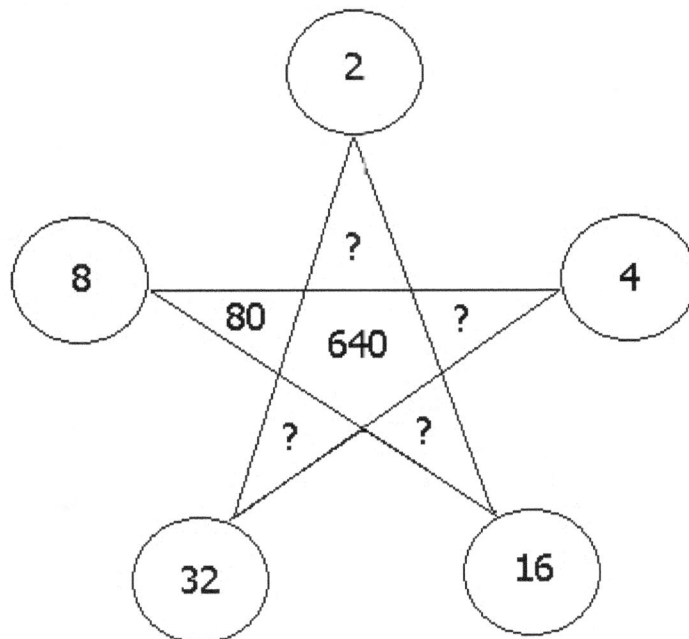

2007 - 2018 © Frank Ho, Amanda Ho, All rights reserved.　　www.homathchess.com

Replace the ? with a number.

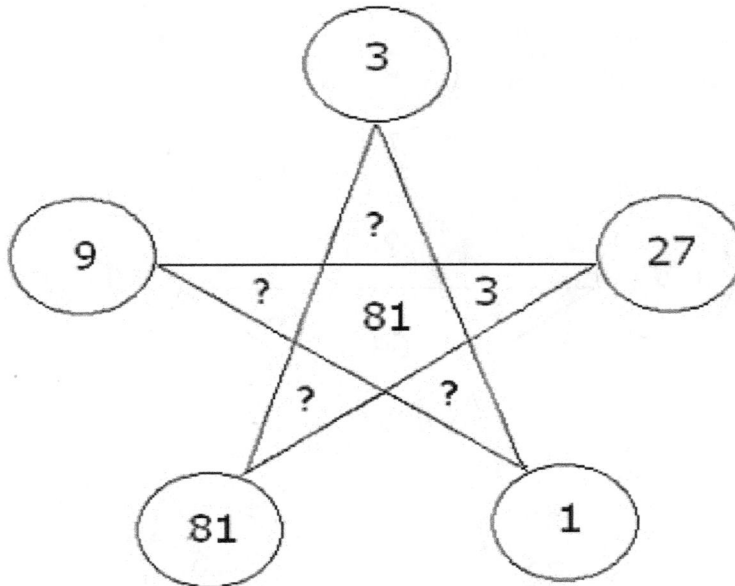

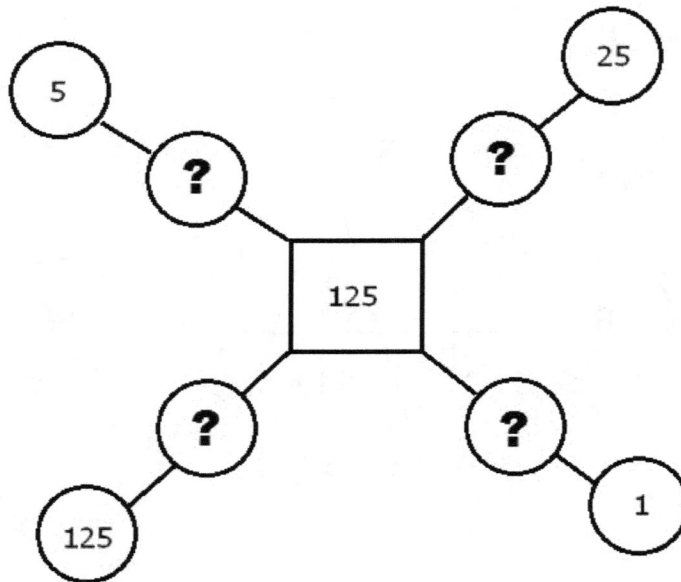

2007 - 2018 © Frank Ho, Amanda Ho, All rights reserved.   www.homathchess.com

Power of 10

$$5 \times 2 = 10$$

$$25 \times 4 = 100$$

$$25 \times 8 = 1000$$

The above strategy can be used in multiplication.

$$5 \times 2 \times 5 \times 2 \times 5 \times 2 = \rule{3cm}{0.4pt}$$

$$25 \times 4 \times 25 \times 4 \times 25 = \rule{2cm}{0.4pt}$$

$$125 \times 8 \times 125 \times 8 \times 125 \times 8 = \rule{3cm}{0.4pt}$$

$$5 \times 2 \times 5 \times 2 \times 5 \times 2 \times 5 \times 2 = \rule{3cm}{0.4pt}$$

$$25 \times 4 \times 25 \times 4 \times 25 \times 4 \times 25 \times 4 \times = \rule{2cm}{0.4pt}$$

$$125 \times 8 \times 125 \times 8 \times 125 \times 8 \times 125 \times 8 = \rule{3cm}{0.4pt}$$

2007 - 2018 © Frank Ho, Amanda Ho, All rights reserved.　　www.homathchess.com

From multiplication to division procedure

In this workbook, the division procedure is based on the reverse procedure of multiplication, For example, $\square \times 2 = 6$, the division procedure is to find out what is the greatest factor of $\square$ such that $\square \times 2 \leq 6$ with the following division notation

$$2\overline{)6}^{\square}$$

There are a few restrictions to follow such as the divisor must be a whole number and the remainder must be less than the divisor and divide the dividends one digit at a time. The reason of dividing the dividend digit one at a time is it automatically takes care of the zeros at the beginning, in the middle, and at the end. The divisor must be whole number would make the decimal division easier.

A simple problem could be used to demonstrate the above concept. For example, $25 is to be divided equally among 5 children. The divisor is 5 children so it must be whole number, but the $ amount could be decimals. If we would write the division as $\$25 \div 5 = \dfrac{\$25}{5}$ then

$$\frac{\$25}{5} = \frac{2\,\$10 + 1\,\$5}{5} = 2\,\$2 + \$1 = \$5$$

If we follow the above procedure for division, it would be very tedious, so we convert 2 $10 to 20 of $1 and add $5 to be 5 of $1. So the division procedure will be

$$5\overline{)25}^{\square}$$. This is the reason that quotient 5 must be placed rightmost to show we are

actually divide 25 of 1$ by 5 children.

For decimal division, the division flowchart included in this workbook can also be used. Bring down 0 until the desired decimal places are found. If the dividend has a decimal point, line up decimal point in the quotient and just carry out division as it is whole number division. If the dividend is a whole number, place decimal point at the place after the dividend digit has been all used and before 0 needs to be brought down.

2007 - 2018 © Frank Ho, Amanda Ho, All rights reserved.　　www.homathchess.com

From multiplication to division (d ÷ d)

Multiplication	Division
Factor × factor $\leq$ product	$divisor \sqrt{dividend}$ with *quotinet* above
$\square \times 2 \leq 6$	$\times \square$ ← step 1: What times 2 is ≤ 6. $2\overline{)6}$ $-\ \square$ ← step 2: 6 − 6 = 0 0 ←Remainder = 0
$\square \times 2 \leq 8$	$\times \square$ ← step 1: What times 2 is ≤ 8. $2\overline{)8}$ $-\ \square$ ← step 2: 8 − 8 = 0 0 ←Remainder = 0
$\square \times 3 \leq$ ♛	$\times \square$ ← step 1: What times 3 is ≤ 9. $3\overline{)9}$ $-\ \square$ ← step 2: 9 − 9 = 0 0 ←Remainder = 0

2007 - 2018 © Frank Ho, Amanda Ho, All rights reserved.   www.homathchess.com

From multiplication to division (d $\div$ d)

$\square \times 2 \leq 8$	$8 = \square \times 2$	$\times \square$ ← step 1: What times 2 is ≤ 8. $2\overline{)8}$ $-\ \square$ ← step 2: $8 - 8 = 0$ 0 ←Remainder = 0
$\square \times 3 \leq 0$	$0 = \square \times$ ♟	$\times \square$ ← step 1: What times 3 is ≤ 0. $3\overline{)0}$ $-\ \square$ ← step 2: $0 - 0 = 0$ 0 ←Remainder = 0
$\square \times 6 \leq 6$	$6 = \square \times 6$	$\times \square$ ← step 1: What times 6 is ≤ 6. $6\overline{)6}$ $-\ \square$ ← step 2: $6 - 6 = 0$ 0 ←Remainder = 0

Ho Math Chess 何数棋谜 妈！我会棋谜式乘法啦！
Mom! I Learn Multiplication Using Math-Chess-Puzzles Connection!

Student's Name _____ Date _____

2007 - 2018 © Frank Ho, Amanda Ho, All rights reserved.　　www.homathchess.com

dd ÷ d with 1-digit quotient and no remainder

☐ × 6 = 42	42 = ☐ × 7	Step1: Do ☐ × 6 ≤ 42 (4 is too small, use 42) **Place the quotient in the rightmost position** × ☐ ← step 2: 7 × 6 = 42 6)‾4‾2‾ −☐☐ ← step 3: 42 − 42 = 0 0 ←Remainder = 0
☐ × 2 = 18	18 = ☐ × ♛	**Step1:** Do ☐ × 2 ≤ 18 (1 is too small, use 18) **Place the quotient in the rightmost position** × ☐ ← step 2: 9 × 2 = 18 2)‾1‾8‾ −☐☐ ← step 3: 18 − 18 = 0 0 ←Remainder = 0
☐ × ♛ = 81	81 = ☐ × 9	**Step1:** Do ☐ × 9 ≤ 81 (8 is too small, use 81) **Place the quotient in the rightmost position** × ☐ ← step 2: 9 × 9 = 81 9)‾8‾1‾ −☐☐ ← step 3: 81 − 81 = 0 0 ←Remainder = 0

2007 - 2018 © Frank Ho, Amanda Ho, All rights reserved.   www.homathchess.com

dd ÷ d with 1-digit quotinet and no remainder

Step1: Do ☐ × 3 ≤ 15 (1 is too small, use 15)

× ☐ ← step 2: Do multiplication, 5 × 3 = 15

3)15

− ☐☐ ← step 3: Do subtraction, 15 − 15 = 0

0 ←Remainder = 0

Step1 Do ☐ × 3 ≤ 18 (3 is too small, use 18)

× ☐ ← step 2: Do multiplication, 6 × 3 = 18

3)18

− ☐☐ ← step 3: Do substraction, 18 − 18 = 0

0 ←Remainder = 0

Step1: Do ☐ × 5 ≤ 25 (2 is too small, use 25)

× ☐ ← step 2: Do multiplication

5)25

− ☐☐ ← step 3: Do subtraction

0 ←Remainder = 0

Step1: Do ☐ × 4 ≤ 28 (2 is too small, use 28)

× ☐ ← step 2: Do multiplication

4)28

− ☐☐ ← step 3: Do subtraction

0 ←Remainder = 0

Step1: Do ☐ × 4 ≤ 32 (3 is too small, use 32)

× ☐ ← step 2: Do multiplication

4)32

− ☐☐ ← step 3: Do subtraction

0 ←Remainder = 0

Step1: Do ☐ × 2 ≤ 16 (1 is too small, use 16)

× ☐ ← step 2: Do multiplication

2)16

− ☐☐ ← step 3: Do subtraction

0 ←Remainder = 0

Mom! I Learn Multiplication Using Math-Chess-Puzzles Connection!

Student's Name _____ Date _____

2007 - 2018 © Frank Ho, Amanda Ho, All rights reserved.   www.homathchess.com

From multiplication to division

$\square \times 2 \leq 12$ $\square \times 6 \leq 12$	$\times\square$ $6\overline{)12}$ $\square\,\square$	$\times\square$ $2\overline{)12}$ $\square\,\square$	$12 \div 2 = \square$ $12 \div 6 = \square$
$\square \times 3 \leq 18$ $\square \times 6 \leq 18$	$\times\square$ $6\overline{)18}$ $\square\,\square$	$\times\square$ $3\overline{)18}$ $\square\,\square$	$18 \div \text{♞} = \square$ $18 \div 6 = \square$
$\square \times 4 \leq 12$ $\square \times 3 \leq 12$	$\times\square$ $4\overline{)12}$ $\square\,\square$	$\times\square$ $3\overline{)12}$ $\square\,\square$	$12 \div 4 = \square$ $12 \div \text{♗} = \square$

Ho Math Chess 何数棋谜 妈!我会棋谜式乘法啦!
Mom! I Learn Multiplication Using Math-Chess-Puzzles Connection!

Student's Name _____ Date _____

2007 - 2018 © Frank Ho, Amanda Ho, All rights reserved.   www.homathchess.com

From multiplication to division

$\square \times 5 \le 15$ $\square \times 3 \le 15$	$5\overline{)15}$ with $\times\square$ above and $\square\square$ below	$3\overline{)15}$ with $\times\square$ above and $\square\square$ below	$15 \div ♗ = \square$ $15 \div ♖ = \square$
$\square \times ♖ \le 30$ $\square \times 6 \le 30$	$6\overline{)30}$ with $\times\square$ above and $\square\square$ below	$5\overline{)30}$ with $\times\square$ above and $\square\square$ below	$30 \div 5 = \square$ $30 \div 6 = \square$
$\square \times 5 \le 40$ $\square \times 8 \le 40$	$5\overline{)40}$ with $\times\square$ above and $\square\square$ below	$8\overline{)40}$ with $\times\square$ above and $\square\square$ below	$40 \div 5 = \square$ $40 \div 8 = \square$

Ho Math Chess   何数棋谜   妈！我会棋谜式乘法啦！
Mom! I Learn Multiplication Using Math-Chess-Puzzles Connection!

Student's Name _____ Date _____

2007 - 2018 © Frank Ho, Amanda Ho, All rights reserved.   www.homathchess.com

From multiplication to division

$\square \times 5 \leq 45$ $\square \times 9 \leq 45$	$5\overline{)45}$ with $\times\square$ above and $\square\square$ below	$9\overline{)45}$ with $\times\square$ above and $\square\square$ below	$45 \div 9 = \square$ $45 \div ♖ = \square$
$\square \times ♖ \leq 10$ $\square \times 2 \leq 10$	$2\overline{)10}$ with $\times\square$ above and $\square\square$ below	$5\overline{)10}$ with $\times\square$ above and $\square\square$ below	$10 \div 5 = \square$ $10 \div 2 = \square$
$\square \times 5 \leq 25$ $\square \times ♖ \leq 25$	$5\overline{)25}$ with $\times\square$ above and $\square\square$ below	$5\overline{)25}$ with $\times\square$ above and $\square\square$ below	$25 \div 5 = \square$ $25 \div 5 = \square$

397

From multiplication to division

$\square \times 4 \leq 20$ $\square \times 5 \leq 20$	$5\overline{)20}$	$4\overline{)20}$	$20 \div 4 = \square$ $20 \div ♖ = \square$
$\square \times ♖ \leq 35$ $\square \times 7 \leq 35$	$7\overline{)35}$	$5\overline{)35}$	$35 \div 5 = \square$ $35 \div 7 = \square$
$\square \times 8 \leq 40$ $\square \times ♖ \leq 40$	$5\overline{)40}$	$8\overline{)40}$	$40 \div ♖ = \square$ $40 \div 8 = \square$

2007 - 2018 © Frank Ho, Amanda Ho, All rights reserved.　　www.homathchess.com

From multiplication to division (d ÷ d)

□ × 4 ≤ 24 □ × 6 ≤ 24	×□ 4)24 □□	×□ 6)24 □□	24 ÷ 4 = □ 24 ÷ 6 = □
□ × 6 ≤ 42 □ × 7 ≤ 42	×□ 7)42 □□	×□ 6)42 □□	42 ÷ 6 = □ 42 ÷ 7 = □
□ × 6 ≤ 36 □ × 6 ≤ 36	×□ 6)36 □□	×□ 6)36 □□	36 ÷ 6 = □ 36 ÷ 6 = □

Ho Math Chess   何数棋谜   妈！我会棋谜式乘法啦！
Mom! I Learn Multiplication Using Math-Chess-Puzzles Connection!

Student's Name _____ Date _____

2007 - 2018 © Frank Ho, Amanda Ho, All rights reserved.   www.homathchess.com

dd ÷ d with 1-digit quotient and no remainder

$\square \times 5 \leq 30$ $\square \times ♖ \leq 30$	$6\overline{)30}$ ×□	$5\overline{)30}$ ×□	$30 \div 5 = \square$ $30 \div 6 = \square$
$\square \times 6 \leq 24$ $\square \times 4 \leq 24$	$4\overline{)24}$ ×□	$6\overline{)24}$ ×□	$24 \div 6 = \square$ $24 \div 4 = \square$
$\square \times ♗ \leq 18$ $\square \times 6 \leq 18$	$6\overline{)18}$ ×□	$3\overline{)18}$ ×□	$18 \div 6 = \square$ $18 \div ♗ = \square$

Student's Name _____ Date _____

2007 - 2018 © Frank Ho, Amanda Ho, All rights reserved.   www.homathchess.com

dd ÷ d with 1-digit quotient and no remainder

$\square \times 5 \leq 45$ $\square \times ♛ \leq 45$	$\times\square$ $5\overline{)45}$ $\underline{\square\square}$	$\times\square$ $9\overline{)45}$ $\underline{\square\square}$	$45 \div 5 = \square$ $45 \div ♛ = \square$
$\square \times ♜ \leq 40$ $\square \times 8 \leq 40$	$\times\square$ $5\overline{)40}$ $\underline{\square\square}$	$\times\square$ $8\overline{)40}$ $\underline{\square\square}$	$40 \div 5 = \square$ $40 \div 8 = \square$
$\square \times 6 \leq 30$ $\square \times 5 \leq 30$	$\times\square$ $6\overline{)30}$ $\underline{\square\square}$	$\times\square$ $5\overline{)30}$ $\underline{\square\square}$	$30 \div 6 = \square$ $30 \div ♜ = \square$

2007 - 2018 © Frank Ho, Amanda Ho, All rights reserved.   www.homathchess.com

dd ÷ d with 1-digit quotient and no remainder

$\square \times ♜ \le 35$ $\square \times 7 \le 35$	$5\overline{)35}$ with $\times\square$ on top and $\square\square$ below	$7\overline{)35}$ with $\times\square$ on top and $\square\square$ below	$35 \div ♜ = \square$ $35 \div 7 = \square$
$\square \times 5 \le 30$ $\square \times 6 \le 30$	$5\overline{)30}$ with $\times\square$ on top and $\square\square$ below	$6\overline{)30}$ with $\times\square$ on top and $\square\square$ below	$30 \div ♜ = \square$ $30 \div 6 = \square$
$\square \times 4 \le 20$ $\square \times ♜ \le 20$	$5\overline{)20}$ with $\times\square$ on top and $\square\square$ below	$4\overline{)20}$ with $\times\square$ on top and $\square\square$ below	$20 \div 4 = \square$ $20 \div ♜ = \square$

2007 - 2018 © Frank Ho, Amanda Ho, All rights reserved.   www.homathchess.com

dd ÷ d with remainder vs. no remainder

$35 \div 7 = \square$	$5\overline{)35}$	$5\overline{)36}$
$20 \div \text{♖} = \square$	$4\overline{)20}$	$4\overline{)22}$
$56 \div 8 = \square$	$7\overline{)56}$	$7\overline{)59}$
$81 \div \text{♕} = \square$	$9\overline{)81}$	$9\overline{)85}$

403

Mom! I Learn Multiplication Using Math-Chess-Puzzles Connection!

Student's Name _____ Date _____

2007 - 2018 © Frank Ho, Amanda Ho, All rights reserved.   www.homathchess.com

dd ÷ d with remainder vs. no remainder

$21 ÷ ♞ = \square$	$\begin{array}{r} ×\square \\ 7\overline{)21} \\ \square\square \end{array}$	$\begin{array}{r} ×\square \\ 7\overline{)25} \\ \square\square \\ \square \end{array}$
$42 ÷ 6 = \square$	$\begin{array}{r} ×\square \\ 7\overline{)42} \\ \square\square \end{array}$	$\begin{array}{r} ×\square \\ 7\overline{)45} \\ \square\square \\ \square \end{array}$
$32 ÷ 4 = \square$	$\begin{array}{r} ×\square \\ 8\overline{)32} \\ \square\square \end{array}$	$\begin{array}{r} ×\square \\ 8\overline{)38} \\ \square\square \\ \square \end{array}$
$30 ÷ 6 = \square$	$\begin{array}{r} ×\square \\ 5\overline{)30} \\ \square\square \end{array}$	$\begin{array}{r} ×\square \\ 5\overline{)33} \\ \square\square \\ \square \end{array}$

2007 - 2018 © Frank Ho, Amanda Ho, All rights reserved.   www.homathchess.com

dd ÷ d with remainder vs. no remainder

$18 \div ♛ = \square$	$\times\square$ $2\overline{)18}$ $\square\square$	$\times\square$ $2\overline{)19}$ $\square\square$ $\square$
$12 \div 6 = \square$	$\times\square$ $2\overline{)12}$ $\square\square$	$\times\square$ $2\overline{)13}$ $\square\square$ $\square$
$14 \div 7 = \square$	$\times\square$ $2\overline{)14}$ $\square\square$	$\times\square$ $2\overline{)15}$ $\square\square$ $\square$
$28 \div 4 = \square$	$\times\square$ $7\overline{)28}$ $\square\square$	$\times\square$ $7\overline{)33}$ $\square\square$ $\square$

Ho Math Chess   何数棋谜   妈！我会棋谜式乘法啦！
Mom! I Learn Multiplication Using Math-Chess-Puzzles Connection!

Student's Name _____ Date _____

2007 - 2018 © Frank Ho, Amanda Ho, All rights reserved.   www.homathchess.com

dd ÷ d with remainder vs. no remainder

$15 \div$ ♖ $= \square$	$\times \square$ $3 \overline{)15}$ $\square\square$	$\times \square$ $3 \overline{)16}$ $\square\square$ $\square$
$18 \div 6 = \square$	$\times \square$ $3 \overline{)18}$ $\square\square$	$\times \square$ $3 \overline{)19}$ $\square\square$ $\square$
$24 \div 8 = \square$	$\times \square$ $3 \overline{)24}$ $\square\square$	$\times \square$ $3 \overline{)26}$ $\square\square$ $\square$
$27 \div$ ♕ $= \square$	$\times \square$ $3 \overline{)27}$ $\square\square$	$\times \square$ $3 \overline{)29}$ $\square\square$ $\square$

2007 - 2018 © Frank Ho, Amanda Ho, All rights reserved.   www.homathchess.com

dd ÷ d with remainder vs. no remainder

$12 \div ♝ = \square$	$4 \overline{)12}$ with $\times\square$ above and $\square\square$ below	$4 \overline{)13}$ with $\times\square$ above and $\square\square$, $\square$ below
$16 \div 4 = \square$	$4 \overline{)16}$ with $\times\square$ above and $\square\square$ below	$4 \overline{)18}$ with $\times\square$ above and $\square\square$, $\square$ below
$20 \div ♜ = \square$	$4 \overline{)20}$ with $\times\square$ above and $\square\square$ below	$4 \overline{)23}$ with $\times\square$ above and $\square\square$, $\square$ below
$24 \div 6 = \square$	$4 \overline{)24}$ with $\times\square$ above and $\square\square$ below	$4 \overline{)27}$ with $\times\square$ above and $\square\square$, $\square$ below

Ho Math Chess 何数棋谜 妈！我会棋谜式乘法啦！
Mom! I Learn Multiplication Using Math-Chess-Puzzles Connection!
Student's Name _____ Date _____
2007 - 2018 © Frank Ho, Amanda Ho, All rights reserved.   www.homathchess.com

dd ÷ d with remainder vs. no remainder

$10 \div 2 = \Box$	$5\overline{)10}$	$5\overline{)14}$
$20 \div 4 = \Box$	$5\overline{)20}$	$5\overline{)24}$
$25 \div ♖ = \Box$	$5\overline{)25}$	$5\overline{)24}$
$40 \div 8 = \Box$	$5\overline{)40}$	$5\overline{)44}$

Mom! I Learn Multiplication Using Math-Chess-Puzzles Connection!

Student's Name _____ Date _____

2007 - 2018 © Frank Ho, Amanda Ho, All rights reserved.   www.homathchess.com

From multiplication to division

Fill in the following ☐ with a number.

$\begin{array}{r} 2 \\ \times\ \underline{\text{♛}} \end{array}$ ☐ ÷ 9 = ☐	$\begin{array}{r} 9 \\ \times\ \underline{2} \end{array}$ ☐ ÷ 2 = ☐
$\begin{array}{r} 3 \\ \times\ \underline{\text{♛}} \end{array}$ ☐ ÷ ♛ = ☐	$\begin{array}{r} \text{♛} \\ \times\ \underline{3} \end{array}$ ☐ ÷ 3 = ☐
$\begin{array}{r} 4 \\ \times\ \underline{9} \end{array}$ ☐ ÷ ♛ = ☐	$\begin{array}{r} 9 \\ \times\ \underline{4} \end{array}$ ☐ ÷ 4 = ☐
$\begin{array}{r} 5 \\ \times\ \underline{\text{♛}} \end{array}$ ☐ ÷ 9 = ☐	$\begin{array}{r} 9 \\ \times\ \underline{5} \end{array}$ ☐ ÷ 5 = ☐
$\begin{array}{r} 6 \\ \times\ \underline{9} \end{array}$ ☐ ÷ ♛ = ☐	$\begin{array}{r} \text{♛} \\ \times\ \underline{6} \end{array}$ ☐ ÷ 6 = ☐
$\begin{array}{r} 7 \\ \times\ \underline{9} \end{array}$ ☐ ÷ 9 = ☐	$\begin{array}{r} \text{♛} \\ \times\ \underline{7} \end{array}$ ☐ ÷ 7 = ☐

2007 - 2018 © Frank Ho, Amanda Ho, All rights reserved.   www.homathchess.com

From multiplication to division

Fill in the following ☐ with a number.

2 $\times\ 8$ ☐ $\div\ 2 =$ ☐	6 $\times\ 2$ ☐ $\div\ 2 =$ ☐
♘ $\times\ 7$ ☐ $\div\ 3 =$ ☐	♖ $\times\ 3$ ☐ $\div\ 3 =$ ☐
4 $\times\ 6$ ☐ $\div\ 6 =$ ☐	4 $\times\ 4$ ☐ $\div\ 4 =$ ☐
♖ $\times\ 5$ ☐ $\div\ 5 =$ ☐	7 $\times\ 5$ ☐ $\div\ $♖$ =$ ☐
6 $\times\ 4$ ☐ $\div\ 4 =$ ☐	8 $\times\ 6$ ☐ $\div\ 8 =$ ☐
7 $\times\ 8$ ☐ $\div\ 7 =$ ☐	6 $\times\ 7$ ☐ $\div\ 6 =$ ☐

Mom! I Learn Multiplication Using Math-Chess-Puzzles Connection!

Student's Name _____ Date _____

2007 - 2018 © Frank Ho, Amanda Ho, All rights reserved.   www.homathchess.com

From multiplication to division

Fill in the following ☐ with a number.

2 × 8 ☐ ÷ 2 = ☐	8 × 2 ☐ ÷ 8 = ☐
4 × 9 ☐ ÷ ♛ = ☐	4 × 3 ☐ ÷ ♞ = ☐
4 × 6 ☐ ÷ 4 = ☐	8 × 4 ☐ ÷ 4 = ☐
5 × 7 ☐ ÷ 7 = ☐	4 × 5 ☐ ÷ 4 = ☐
6 × 8 ☐ ÷ 6 = ☐	6 × 6 ☐ ÷ 6 = ☐
7 × ♞ ☐ ÷ 3 = ☐	4 × 7 ☐ ÷ 7 = ☐

Mom! I Learn Multiplication Using Math-Chess-Puzzles Connection!

Student's Name _____ Date _____

2007 - 2018 © Frank Ho, Amanda Ho, All rights reserved.   www.homathchess.com

From multiplication to division

Fill in the following ☐ with a number.

☐   ☐

$\dfrac{}{\times\ 3} \quad \div\ \dfrac{3}{9}$
♛

☐   ☐

$\dfrac{}{\times\ ♜} \quad \div\ \dfrac{5}{6}$
6

☐   ☐

$\dfrac{}{\times\ 5} \quad \div\ \dfrac{5}{♛}$
9

☐   ☐

$\dfrac{}{\times\ 7} \quad \div\ \dfrac{7}{6}$
6

☐   ☐

$\dfrac{}{\times\ 5} \quad \div\ \dfrac{♜}{7}$
7

☐   ☐

$\dfrac{}{\times\ 8} \quad \div\ \dfrac{8}{9}$
♛

Ho Math Chess   何数棋谜   妈！我会棋谜式乘法啦！
Mom! I Learn Multiplication Using Math-Chess-Puzzles Connection!

Student's Name _____ Date _____

2007 - 2018 © Frank Ho, Amanda Ho, All rights reserved.   www.homathchess.com

From multiplication to division

Fill in the following ☐ with a number.

☐ ☐ ___×__5 ÷ ♖ 9 ♕	☐ ☐ ___×__8 ÷ _8_ 6 6
☐ ☐ ___×__♖ ÷ _5_ 8 8	☐ ☐ ___×__4 ÷ _4_ 6 6
☐ ☐ ___×__6 ÷ _6_ 7 7	☐ ☐ ___×__8 ÷ _8_ ♗ 3

Ho Math Chess   何数棋谜   妈！我会棋谜式乘法啦！
Mom! I Learn Multiplication Using Math-Chess-Puzzles Connection!

Student's Name _____ Date _____

2007 - 2018 © Frank Ho, Amanda Ho, All rights reserved.   www.homathchess.com

From multiplication to division

Fill in the following ☐ with a number.

2007 - 2018 © Frank Ho, Amanda Ho, All rights reserved.　　www.homathchess.com

From multiplication to division

Fill in the following ☐ with a number.

Mom! I Learn Multiplication Using Math-Chess-Puzzles Connection!

Student's Name _____ Date _____

2007 - 2018 © Frank Ho, Amanda Ho, All rights reserved.   www.homathchess.com

From multiplication to division

Fill in the following ☐ with a number.

☐ ☐ _____ ÷ ♖ × 5 6 6	☐ ☐ _____ ÷ 6 × 6 ♛ 9
☐ ☐ _____ ÷ 6 × 6 8 8	☐ ☐ _____ ÷ 7 × 7 8 8
☐ ☐ _____ ÷ 7 × 6 6 7	☐ ☐ _____ ÷ 6 × 6 4 4

Ho Math Chess　　何数棋谜　妈!我会棋谜式乘法啦!
Mom! I Learn Multiplication Using Math-Chess-Puzzles Connection!
Student's Name _____ Date _____
2007 - 2018 © Frank Ho, Amanda Ho, All rights reserved.　　www.homathchess.com

介紹何数棋谜

何数棋谜=奧数棋谜　+ 思唯腦力開發
英文教材, 中英双语教学

什麼是何数棋谜?

上百篇科學論文巳發表國際象棋可以提高兒童問題解答能力. 並且訓練他們的專心及耐力. 所以我們已經知道下國際象棋對兒童有好處. 但是因為國際象棋與計算能力並無直接開係,所以如何讓兒童能在一個歡樂的環境下也能利用下棋來提高數學的計算呢? 何老師首創並發明有版权的幾何棋藝符號並利用此符號發明了世界第一的独特结合數學与棋謎教材. 何数棋谜讓兒童能利用幾何棋藝符號進行邏輯推理及數字的運算. 棋藝與算術的綜合題含蓋了整數,幾何,集合,抽象數,對比異同,函數,座標,多空間圖形資料,及規則性數字分析. 並且把棋藝的趣味性和數學的知識性結合在一起.

何数棋谜如何幫助兒童腦力思唯的開發?

很簡單的一個道理就是讓學生自願地去用腦,何数棋谜首創獨一無二的融合數學與棋谜的独特趣味寓教於樂教材,利用國際象棋訓練右腦的座標,空間分析及圖形處理,並利用發明了整合棋子與數學的圖形語言,讓兒童能利用符號圖形訓練左腦進行邏輯推理及數字的運算. 國際象棋與算術的綜合題含蓋了整數,幾何,集合,抽象數,對比異同,函數,多空間圖形資料. 所以枯燥無味的計算題變成了謎題,學生需要通過更多的思考. 能讓腦去思考愈多則腦力也愈開發. 處里訊息,分析資料才能發掘出題目. 做這些謎題式數學時可以训練學生比較會專心及有耐心.

2007 - 2018 © Frank Ho, Amanda Ho, All rights reserved.   www.homathchess.com

何**数棋谜**融合數學與國際象棋的教學理論已在 BC 省數學教師刊物上發表. 科研報告已經證實何**数棋谜**教學法不但可以提高兒童數學解題及思維能力,還可以開發兒童的腦力,及分析問題的能力並且增加兒童學習的耐力,學生的探索創造精神及求知欲. 判斷力,及自信心等,啓發思維訓練機警靈巧及加強手腦眼的靈活運用.

2007 - 2018 © Frank Ho, Amanda Ho, All rights reserved.   www.homathchess.com

Introducing Ho Math Chess™

Ho Math Chess™ = math + puzzles + chess

Frank Ho, a Canadian math teacher, intrigued by the relationships between math and chess after teaching his son chess started **Ho Math Chess™** in 1995. His long-term devotion of research has led his son to become a FIDE chess master and Frank's publications of over 20 math workbooks. Today **Ho Math Chess™** is the world largest and the only franchised scholastic math, chess and puzzles specialty learning center with worldwide locations. **Ho Math Chess™** is a leading research organization in the field of math, chess, and puzzles integrated teaching methodology.

There are hundreds of articles already published showing chess benefits children and that math puzzles are a very good way of improving brainpower. So, by integrating chess and mathematical chess puzzles together, the learning effect is more significant.

Parents send their children to **Ho Math Chess™** because they like **Ho Math Chess™** teaching philosophy – offering children problem-solving questions in a variety of formats. The questions could be pure chess, chess puzzles or mathematical chess puzzles in nature of logic, pattern, tree structure, Venn diagram, probability and many more math concepts.

Ho Math Chess™ has developed a series of unique and high quality math, chess, and puzzles integrated workbooks. **Ho Math Chess™** produced the world's first workbook **Learning Chess to Improve Math.** This workbook is not only for learning chess, but also for enriching math ability. This sets **Ho Math Chess** apart from other math learning centers, chess club, or chess classes.

The teaching method at **Ho Math Chess™** is to use math, chess, and puzzles integrated workbooks to teach children fun math. The purposes of **Ho Math Chess™** teaching method and workbooks are to:

- Improve math marks.
- Develop problem solving and critical thinking skills.
- Improve logic thinking ability.
- Boost brainpower.

Testimonials, sample worksheets, reports, and franchise information can be found at www.homathchess.com.

More information about **Ho Math Chess™** can also be found from the following publications:

1. Why Buy a **Ho Math Chess™** Learning Centre Franchise: A Unique Learning Centre?
2. **Ho Math Chess™** Sudoku Puzzles Sample Worksheets
3. Introduction to **Ho Math Chess™** and its Founder Frank Ho

The above publications can be purchased from www.amazon.com.

www.ingramcontent.com/pod-product-compliance
Lightning Source LLC
Chambersburg PA
CBHW082126210326
41599CB00031B/5890